FLOWERS OF THE BRECKS

PART TWO: WETLAND, WOODLAND & FARMLAND

By **Mike Crewe**

Published by the British Trust for Ornithology

About the Brecks

As you travel across East Anglia and enter the area known as the Brecks, the look and feel of the landscape changes dramatically. Wide vistas, vast skies and sandy fields divided by rows of twisted pines all confirm a sense that this is a world apart. Also known as Breckland, it covers some 370 square miles of inland Norfolk and Suffolk, extending roughly north to south from Swaffham to Bury St Edmunds, and west to east from Lakenheath to East Harling. Characterised by generally sandy soils with layers of chalk and flint, and by a climate that is among the driest in Britain, the local landscapes have a distinctive quality. Native forest cover was largely cleared by the Neolithic period, some 4,500 years ago, giving way to a largely open steppe-type habitat of heathland and with swathes of inland sand dunes.

Away from the river valleys, the soil was light and nutrient-poor, with agriculture basic and often temporary. Sections of heath were periodically ploughed or broken (hence the term "brecks") and then cropped for a few years before the land was exhausted and the fields allowed to revert to heathland. Sheep husbandry and the rearing of Rabbits in managed warrens proved to be more productive forms of land use, ensuring that the treeless open heaths of the area survived until the early years of the 20th century. Only with the arrival of large-scale commercial timber production (the Forestry Commission began planting Thetford Forest in 1922) and intensive arable farming, made possible by artificial fertilisers and pesticides, did the traditional Brecks landscape begin to change.

Wildlife abounded on the open heaths, which became the last refuge of species driven to the brink of extinction elsewhere in Britain. It was a world described lyrically by W G Clarke – a local amateur archaeologist and natural historian who coined the term "Breckland" – in his many articles and celebrated book, *In Breckland Wilds* (1925). Although the face of the Brecks has changed much since Clarke's day, the area remains rich in wildlife. The forestry plantations have provided a range of new habitat opportunities, and recent heathland restoration schemes have helped improve the fortunes of many of the specialist birds and plants. A biodiversity audit published by the University of East Anglia in 2010 revealed that the area supports some 12,500 species, including 28 per cent of all those considered rare or under threat – more than in any other part of the UK. With several flagship reserves and many other accessible habitats, the Brecks offers some of the best wildlife watching anywhere in the country.

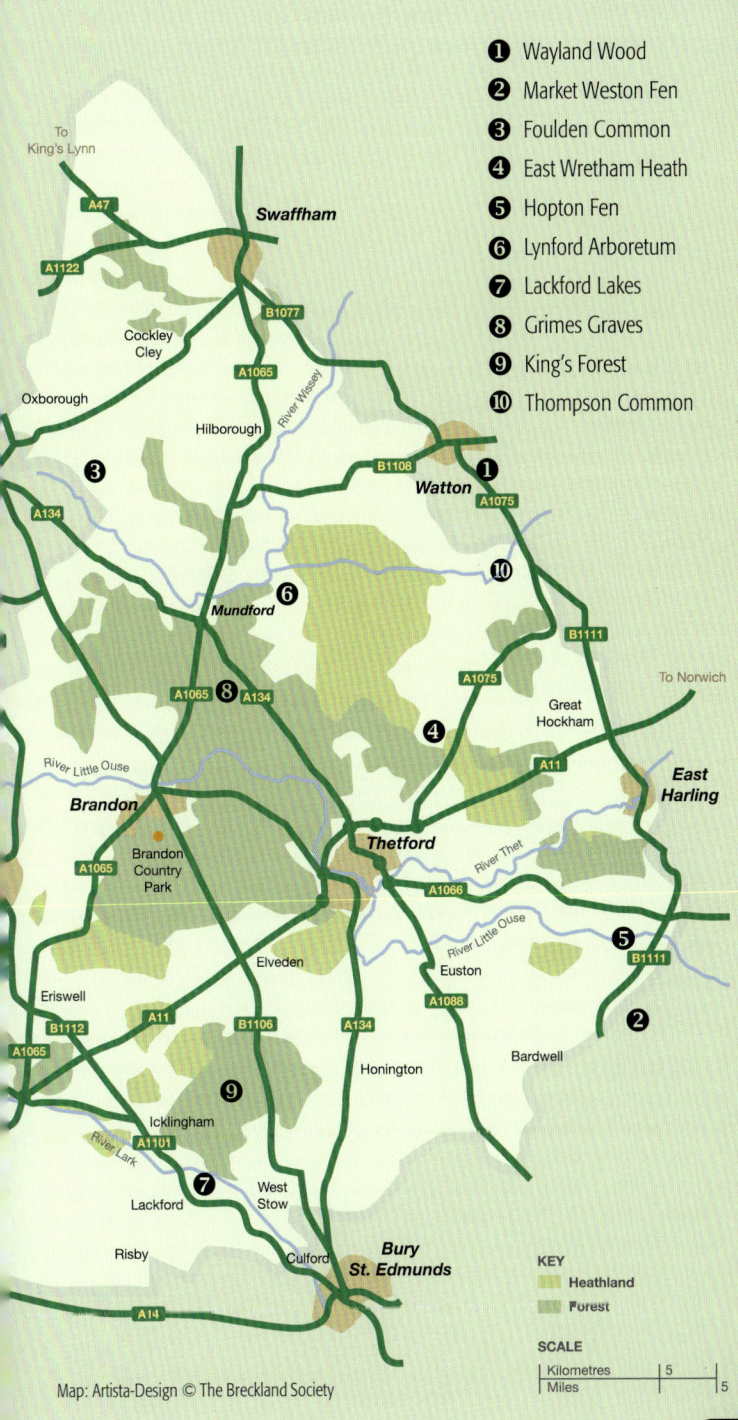

Habitats: Wetlands

While Breckland was once overwhelmingly dominated by grassy heaths and open sandy places, much of it has since been cultivated to the extent that arable farmland and plantation forest are now important habitats. This second book on the flowers of the Brecks covers the plants of these manmade landscapes, as well as those found along Breckland's major waterways and around its lakes and ponds. At the end of the book are a few species that grow on stone walls in the area.

Rivers & Ponds

Because of the particularly free-draining nature of Breckland's soils, surface water is relatively scarce. However, the watersheds of the Rivers Lark, Little Ouse, Thet and Wissey support a wonderful array of wetland plant species. These rivers are too small or silted-up for extensive navigation and are generally shallow, providing plenty of opportunity for wetland vegetation to develop. Some sections are deeper and slower-moving, with stands of taller fen habitat along their banks, while shallow, stony sections can be a riot of white Water-crowfoot in summer.

Many of Breckland's ponds have an interesting origin; known as pingos, they began life as ice-filled depressions created by glacial activity and some are now filled with water more or less all year round. The area's celebrated meres are seasonally flooded in some years but remain dry in others, for reasons that remain not fully understood. Such variability produces a wonderful range of habitats for flora including everything from true aquatics, such as pondweeds and water-lilies, to terrestrial plants that appear as waifs for short periods during spells when the ponds are dry.

Where to visit (see map): The Lark Valley Path offers great river access from Mildenhall to Icklingham, as does the Little Ouse Path from Thetford to Brandon. Lackford Lakes Nature Reserve (Suffolk Wildlife Trust) has well-vegetated, former gravel extraction pits, while the area around Thompson Common (Norfolk Wildlife Trust) hosts a spectacular set of well-managed pingos. East Wretham Heath (also NWT) has two important meres, Langmere and Ringmere.

Marshy Fens and Bogs

Because of the unusual geology of Breckland, where acidic sands lie in shallow drifts over a bedrock of chalk, both calcareous fens and acidic bogs can sometimes be found close to each other, providing some excellent plant-hunting. Valley fens along the Little Ouse may hold dense sedge beds with tall stands of Hemp-agrimony, Great Willowherb and other wetland plants at their edges, while secluded mats of Sphagnum and stands of Blunt-flowered Rush provide homes for various marsh orchids.

Where to visit (see map): Water meadows can be found at the Norah Hanbury-Kelk Meadows at Mildenhall (SWT, with restricted access to the reserve but some nice areas nearby) and interesting marshy areas exist at Foulden Common and Thompson Common. Flower-rich acidic bogs, as well as some chalky fen habitats, can be enjoyed at Hopton and Market Weston Fens (SWT).

Habitats: Woodlands

Deciduous Woodland

The shallow, dry soils of much of Breckland are not suited to the development of natural woodland and it is mainly along the boundaries of the area, on deeper boulder clay, where one can find coppiced woodland with an exciting ground flora of Bluebells, Wood Anemones and even the wonderful Yellow Star-of-Bethlehem. However, deciduous belts of Beech, mixed with Sycamore and various limes, were planted during the late 18th and early 19th centuries and some of these are now home to patches of woodland flora, including some good colonies of Broad-leaved Helleborine. Wet alder or willow carr occurs in places along the main watercourses, but can be difficult to explore.

Where to visit (see map): Wayland Wood (NWT) offers probably the best woodland flora in the area, but deciduous plantations and mixed woodland around The King's Forest, Lynford and Santon Downham (Forestry Commission) are all worth exploring for wildflowers.

Coniferous Woodland

Perhaps the greatest change to take place in the recent landscape history of Breckland has been the extensive planting of pines and other conifers by the Forestry Commission, mostly during the first half of the 20th century. Known collectively as Thetford Forest, the plantations cover vast tracts of what was once heathland or degraded grazing land, stretching right across the area and even beyond, into the heaths of northwest Norfolk. These woods are rather poor when it comes to variety of plants, offering little else than great swathes of bracken or bramble below the trees. However, open paths and rides through the forests reveal remnants of former heathland and there are plenty of surprises to look out for, including Purple Milk-vetch and Barberry in grassy places and Smooth Rupturewort on sandy tracks.

Where to visit (see map): Forest tracks in the Grime's Graves area are especially interesting to explore, as are those around Cranwich and Weeting, and close to Barton Mills and Mildenhall.

Habitats: Arable

Arable & Disturbed Land

The enclosure of Breckland's heaths gradually saw vast tracts of land put under the plough and today much of the area is given over to the production of cereals, root crops and livestock (especially pigs and poultry). While this has changed the landscape dramatically, the rows of iconic twisted pines surrounding neat green croplands and pig fields provide sandy verges and rough corners that are home to a great number of both native and introduced flowers, including some of the Breckland specialities. While modern agricultural practice tends to have a negative impact on our flora, collaborative initiatives between conservation organisations and local farmers can be effective at improving prospects for some of the rarer plants, with Breckland, Spring and Fingered Speedwells all benefiting from such schemes.

Where to visit: Croplands are private and should not be entered, although arable margins can be investigated along the sides of roads, footpaths and permissive tracks. Disturbed ground in Thetford and Brandon may also turn up some interesting plants, including the rarer speedwells and Smooth Rupturewort.

Recording, monitoring and learning more

Making your sightings count – why should I record wildlife?
Being out in the countryside and identifying the wildflowers that you find is always exciting, but the value of your enjoyment can be greatly enhanced when your sightings are shared with organisations that seek to understand our wildlife. Knowing what wildflowers are present on a site that you visit is really important, and the building block to knowing how an area can be managed to conserve wildlife. Noting down or 'recording' your sightings of wildflowers as 'biological records' is not only vital for managing our best wildflower sites, but it also allows us to protect those sites from the impact of housing and other types of development.

What should you record?
A biological record is simply a note of a species observed by a person at a location on a given date. Records must contain four essential pieces of information:

- What species was observed (either common name or scientific name).
- Where was it observed (ideally a six-figure grid reference or better).
- When was it observed (ideally the exact date).
- Who recorded it (full name of the person who made the sighting).

Who should you send records to?
The Norfolk Biodiversity Information Service (NBIS) and Suffolk Biodiversity Information Service (SBIS) are the Local Environmental Records Centres (LERC) covering the Brecks. These organisations are the central point for all records in their respective counties. You may submit your records by using their online systems or you can email the records directly. As you become more interested in recording wildflowers you may want to send your records direct to the relevant County Recorder, who verifies the records NBIS and SBIS receive. Information on who these people are can be found on the Norfolk and Norwich Naturalists' Society (NNNS) and the Suffolk Naturalists' Society (SNS) websites. If you are recording in more than one part of the country you could record on one platform, the most popular of which is iRecord (website and app).

Getting involved in monitoring and surveys
If you are reasonably competent at recording wildflowers and/or wish to volunteer more regularly you can take part in many monitoring or survey projects.
A good place to start is the Botanical Society of Britain & Ireland (BSBI). The National Plant Monitoring Scheme is an annual habitat-based plant monitoring scheme showing changes in our wild plants and their habitats.

Learning more
If you are new to plant hunting, or would like to further your identification skills, the following are good ways to learn and get involved:

- There are many local clubs and societies that will put you in touch with like-minded people, such as the NNNS or SNS (see above).
- The Identiplant programme is an online correspondence course run jointly by BSBI and the Field Studies Council and has proven to be a wonderful way to learn plant identification.
- Ispot is a friendly and free community helping you to identify and share wildlife sightings and is a great place for beginners (www.ispotnature.org).

Star species

Marsh Helleborine
Epipactis palustris

One of our most spectacular wildflowers, with an exotic look more like the kind of orchid you might find in a tropical rainforest. Yet this is a native species, found on a handful of preserved wetlands along the Little Ouse and the Wissey, where it favours species-rich fen habitat with chalky groundwater. It can be recognised early in the season by its tapered but broad-based leaves, which clasp the stem at their base and have well-defined parallel veins. The flowers are at their best in late June and July – a little later than most of our orchids – and are rather variable, with some plants having much denser heads of flowers than others. The petals are usually a combination of white and dark flesh-pink, but rarely a plant may be found where the pink is replaced by pale yellow.

This species was lost from many wet meadows and fens during post-WW2 land improvements and drainage, but it still does well in its remaining strongholds and may form quite dense colonies. Odd plants may yet be found at new sites, so it is always worth keeping a sharp eye out along the main river valleys. Fragile roots mean plants are easily damaged by trampling, so care should always be taken – use a long lens for photography unless boardwalks etc allow a closer approach.

Similar species: Marsh Helleborine is very distinctive and should be readily recognisable by its flowers. Broad-leaved Helleborine (page 40) is somewhat similar, but it has greener flowers and is a plant of drier and usually shady places.

Wetland

Pale Marsh Orchid
Dactylorhiza incarnata ssp. ochroleuca

This plant is so rare that it is unlikely to be found by the casual visitor to the Brecks, but it is a real speciality of the area and deserves its place in this book. Found in species-rich, wet fen habitats with chalky groundwater, the dense spikes of unspotted, cream-coloured flowers are at their best in late May and early June. The leaves are unspotted, sword-like and pale green in colour.

Pale Marsh Orchid seems never to have been common in the UK and is now confined to just three sites in East Anglia. Two of these, Hopton Fen and Market Weston Fen, lie just within the boundary of Breckland in the valley of the Little Ouse and are protected reserves, while the third location at Chippenham Fen in Cambridgeshire is only just outside our area. This species is listed as Critically Endangered in the UK and its growing requirements seem so restricted to a precise type of wet – and yet not permanently wet – fen habitat that its future seems highly uncertain.

Similar species: White-flowered forms of other marsh orchid species (see pages 12–13) and subspecies are occasionally found, but Pale Marsh Orchid can be told from these by its creamy yellow flowers, the clearly three-lobed lower lip and the long greenish bracts in the flowerhead.

White Water-lily
Nymphaea alba

An uncommon species in the Brecks, found most frequently along the Rivers Wissey and Lark and occasionally elsewhere. The flowers of native plants are pure white and relatively small (though sometimes up to 20 cm across) compared with introduced water-lilies. The leaves are circular and float on the water surface in permanent ponds and slow-flowing rivers.

Similar species: Ornamental water-lilies, which are sometimes introduced to ponds and lakes, have flowers that may be pink- or cream-coloured and leaves often spotted with purple.

Yellow Water-lily
Nuphar lutea

Widespread and often forming large colonies in slow-moving or still, permanent waterbodies. The rich-yellow, petal-like sepals fade to leave a flask-like central seed capsule which earns the plant the old name of Brandy-bottle. Leaves are less circular and more oval in outline than those of White Water-lily. Plants usually have large quantities of submerged, rather lettuce-like lower leaves, which are often apparent even when floating leaves are absent.

Yellow Iris
Iris pseudacorus

With their long, sword-like leaves and showy displays of rich yellow flowers, Yellow Irises are a fine sight when in full flower. They grow to a metre or more in height and can form extensive stands in many kinds of wetland areas, but especially around the margins of muddy ponds, ditch sides in fens and in shallow spots along the main river systems. The flowers are easily recognised by their three broad outer petals that hang like flags, and the three narrower inner petals that stand upright in the centre of the flower.

Wetland

Arrowhead
Sagittaria sagittifolia

A plant of permanent watercourses and found commonly along smaller rivers and streams. It grows submerged in muddy rivers and has three types of leaves: those underwater are linear in shape, while those floating on the surface are elliptical and emergent leaves are arrow-shaped. This species is something of a shy flowerer with the leaves often noticed with no flowers present. The flowers have three white petals with a deep maroon-red centre and are carried in small whorls on spikes that appear poking straight out of the water.

Common Water-plantain
Alisma plantago-aquatica

Found in both permanent and temporary wetlands, Common Water-plantain is often one of the first plants to appear on bare mud in disturbed wetland areas, even where livestock have puddled the ground. The three-petalled flowers are small (about 1 cm across) and borne in large numbers on a spreading, many-branched, leafless flower spike that may reach nearly a metre in height. The leaves are all basal and rather dock-like, but with heart-shaped bases and parallel veins.

Grass-of-Parnassus
Parnassia palustris

A real gem of a plant, but sadly rare throughout East Anglia. In Breckland it hangs on in small numbers at protected wetland habitats around Foulden and Market Weston. The attractive five-petalled flowers are rather like white buttercups, but have distinctly prominent veins. The narrowly heart-shaped leaves are yellowish-green in colour and form a cluster at the base of the plant. This species requires peaty soils with chalky groundwater and has declined greatly due to wetland drainage and land improvement schemes.

Bogbean
Menyanthes trifoliata

A showy and easily recognised species that may be found in very wet fens (including in open water) or around the edges of pingos in the upper reaches of the area's water catchments, especially the Wissey and the Lark and in fens along the Little Ouse. Deep reddish-pink buds open to white flowers, with each petal bearing a distinctive covering of white fringes. The leaves are three-lobed, like giant clover leaves, and appear above the water surface at the ends of creeping stems.

Marsh Fragrant Orchid
Gymnadenia densiflora

A rare species of wet fens, this beautiful orchid can still be found at protected sites around Foulden and Market Weston, but was formerly more widespread. The cerise-pink flowers are clustered together in a narrow spike and have very long nectar spurs at the back of the flower, a feature that distinguishes them from the marsh orchids. Leaves are green and unspotted.

Similar species: Chalk Fragrant Orchid (*Gymnadenia conopsea*) is probably now extinct in Breckland, but it occurs in chalk grassland elsewhere in East Anglia.

Heath Spotted Orchid
Dactylorhiza maculata

A species that favours acidic wetlands and is therefore rare in Breckland because of the area's chalky groundwater. Most likely to be found in patches of Sphagnum bog in the upper reaches of the Little Ouse. Flower spikes appear in early summer and can be very variable in colour, some bearing deep pink flowers with heavily spotted petals, with others white and almost unmarked. Leaves are dark-spotted.

Similar species: Common Spotted Orchid (page 40) is very similar, but has deeply three-lobed lower lip to flower.

Wetland

Southern Marsh Orchid
Dactylorhiza praetermissa

The commonest of the wetland orchids in the Breckland water catchments and quite widespread in wet meadows and fens. The flowers have a broad three-lobed lower lip and are usually a rich deep pink, heavily marked with darker spots and lines. The two side petals spread out widely to the side. Leaves are usually all green, but vary and may have dark ring-spots in some colour forms. Often appears in colonies, with many plants packed into a small area.

Early Marsh Orchid
Dactylorhiza incarnata

This species has a number of different forms, but most of the plants found in Breckland are of the subspecies *incarnata*. This form has flower spikes that are typically relatively short and stocky, with flowers that are flesh-pink rather than bright pink. The flowers are relatively narrow, with the lower lip distinctly folded back and the two side petals tending to project upwards rather than outwards (compare with Southern Marsh Orchid). Leaves all green without spots. Found in damp sites along most of the Breckland rivers.

Himalayan Balsam
Impatiens glandulifera

A very variable plant that can be tall and imposing, reaching over two metres in height. The pink, red or white flowers have a unique helmet-like appearance and appear in large showy heads at the top of the stems. Leaves are opposite, linear and have toothed margins. A highly invasive introduced species that is colonising wetlands throughout the UK, thanks in part to its exploding seedpods, although it currently remains rather scarce in Breckland.

Marsh-marigold
Caltha palustris

A very showy relative of the buttercups that forms rounded, leafy mounds in muddy fens and ponds and also in seasonally wet woodland. The flowers are up to 45 mm across and carried in many-branched heads above the leaves. The leaves are heart-shaped and continue to grow larger after the flowers have passed. This is one of the earlier flowers to come out in spring, with those in shady places flowering before the trees have fully leafed up.

Greater Spearwort
Ranunculus lingua

A very tall buttercup, often growing to a metre or more in height. Flowers large for a buttercup (up to 5 cm across) and carried on long stems. Leaves up to 25 cm long and narrowly lance-shaped. Generally a rare plant in the Brecks, but locally frequent in wetlands around Great Hockham. Although a native species, Greater Spearwort is also often cultivated and may appear in village ponds and similar places as a result of deliberate plantings.

Lesser Spearwort
Ranunculus flammula

Appearing like a much smaller version of Greater Spearwort, this species can be quite common in neutral or acidic wet heaths and bogs and may form quite large colonies. The leaves are narrowly lance-shaped but much smaller than those of its larger cousin and it is a low, trailing plant. The flowers are like those of buttercups, but the petals are narrower, with clear gaps between them.

Similar species: Compare with other buttercup species, most of which have broader leaves and grow in drier places.

Wetland

Common Water-crowfoot
Ranunculus aquatilis

The water-crowfoots are essentially aquatic buttercups with white, yellow-centred flowers. Common Water-crowfoot is a plant of still or slow-moving water in ponds and ditches. It usually has broadly three-lobed floating leaves, while the submerged leaves are deeply cut to short bundles of thread-like segments. Plants may be floating in water or growing from damp mud, in the latter case being only of the three-lobed leaf type.

River Water-crowfoot
Ranunculus fluitans

A plant of running water on chalky substrates that can be found growing in the Rivers Lark, Little Ouse and Wissey. A large species that can grow stems up to 6 m or so in length and which may form long tresses in the water, rooted in the bottom and trailing in the current. Leaves are all submerged and reduced to long, thread-like bundles. Flowers poke above the water surface in summer and form islands of colour in the main river channels.

Thread-leaved Water-crowfoot
Ranunculus trichophyllus

Very similar to River Water-crowfoot but generally a smaller plant of still or slow-moving water. Grows submerged; thread-like leaves.

Similar species: All water-crowfoots are superficially rather similar and can be difficult to identify specifically. On the yellow base of each petal there is a small nectar pit (hand lens needed!) that is round in Common, pear-shaped in River and crescent-shaped in Thread-leaved. Three other species – Fan-leaved, Pond and Stream Water-crowfoots – also occur, but a detailed flora is recommended for these.

Celery-leaved Buttercup
Ranunculus sceleratus

A very distinctive annual plant of all kinds of wet places and often found as one of the first colonisers of bare muddy places. The three-lobed basal leaves are very similar to those of celery (hence the name), while stem leaves are narrower. The smooth and shiny stems grow to around 60 cm in height and are topped with odd-looking flowers whose petals appear to be too small for the chunky green centre. Commonly found in disturbed areas, such as where livestock come to drink.

Yellow Loosestrife
Lysimachia vulgaris

A tall plant, growing to a metre or more in height. Leaves lance-shaped and may be in pairs or whorls of three. Flowers appear towards top of the stem in small clusters in the leaf axils. Rare in the Suffolk Brecks, but more frequent in Norfolk along main river valleys and wetlands.

Similar species: Dotted Loosestrife (*L. punctata*) is grown as a garden plant and may be found on roadsides or waste places. It is very similar to Yellow Loosestrife, but forms dense clumps and has larger clusters of flowers.

Greater Bird's-foot-trefoil
Lotus pedunculatus

The tight clusters of pea-like flowers, carried on long flower stems, are a familiar sight in many of the area's wetlands. The plant scrambles with long, hairy stems to 70 cm or more, through tall grasses and other vegetation, and has three-lobed leaves.

Similar species: Common Bird's-foot-trefoil (see **Part One**) is very similar, but is a much smaller species and grows in drier habitats. Greater Bird's-foot-trefoil differs in having spreading sepals, is more hairy and has hollow stems.

Wetland

Strawberry Clover
Trifolium fragiferum

A low-growing, patch-forming species with three-lobed clover leaves that lack the pale marks of similar clovers. The narrow flowers are pale pink with darker veins. After flowering, the seed-bearing heads swell into reddish, rounded balls and look a little like pale strawberries. Although largely a plant of coastal marshes in East Anglia, Strawberry Clover may be found in one or two damp grassland sites in Breckland.

Similar species: Red and White Clovers (see Part One) are more common; they do not form strawberry-like seedheads.

Common Meadow-rue
Thalictrum flavum

A tall and showy plant which may grow to over a metre in height and can be found in tall vegetation in damp meadows and fens beside rivers and streams. The leaves are divided into many small leaflets, which have obvious veins and sharply pointed lobes. The flowers have no petals but a single row of small petal-like sepals from which emerges a tassel of white, creamy-tipped stamens, forming a frothy moptop of flowers in a many-branched head.

Meadowsweet
Filipendula ulmaria

One of the joys of summer is the sweet smell of the flowers of this plant on a warm evening. Very common, growing to a metre in height and found in all kinds of damp places, including roadside ditches. The five-petalled flowers are individually tiny but borne in large frothy masses, held above the coarsely pinnate leaves.

Similar species: Dropwort (see Part One) is a similar but much smaller plant, found on dry, chalky grass heaths.

Purple Loosestrife
Lythrum salicaria

A colourful and attractive plant with long, tall spikes of rich pinkish-purple flowers. These are carried in whorled clusters and have six narrow and rather irregularly sized petals. Leaves are lance-shaped, hairy and rather distinctly veined, with the side veins curving into each other to run parallel with the midrib, rather than reaching to the outer edge of the leaf. A common plant, found along the main river valleys in a wide range of wetland plant communities.

Common Water-pepper
Persicaria hydropiper

The thin, graceful stems of this annual species grow to about 40 cm in height and might easily be missed among other vegetation. The leaves are long, narrow and willow-like, and the tiny white flowers are borne in long, arching spikes at the ends of the stems. The plant derives its English name from its peppery taste. More common in the northern part of Breckland, where it is most often found in damp muddy areas along tracks and streams where they pass through woodland or shady places.

Amphibious Bistort
Persicaria amphibia

A creeping perennial with prostrate stems that can form quite extensive patches. Tight, upright clusters of bright pink flowers appear at the upturned ends of the shoots. An unusual plant in having two distinct growth forms. Terrestrial plants are rather erect and grow in damp grassy meadows and on riverbanks. Aquatic forms can be found along the edges of permanent ponds, with the stems growing out into the water and the leaves floating on the surface.

Wetland

Great Willowherb
Epilobium hirsutum

Growing to a metre and a half in height, large stands of Great Willowherb are an impressive sight in summer. The bright pink flowers are much larger (up to 20 mm across) than those of the other wetland willowherbs and the whole plant is softly hairy. This is a very common species that grows in most kinds of damp places, but is especially frequent in disturbed muddy places such as roadside drainage ditches and can also occur as an urban weed.

Marsh Willowherb
Epilobium palustre

A more distinctive member of a tricky group, this species is more or less hairless and has rounded stems without ridges. Small flowers appear at the top of the plant on elongated stalks and often arch gracefully to one side. It does not typically turn up as a weed in the way that other willowherbs do. The stigma is club-shaped, not four-lobed.

Similar species: Other willowherbs occur in wetlands and can be difficult to tell apart (see page 75). Hoary Willowherb is very similar, but has a dense covering of whitish hairs.

Ragged-robin
Silene flos-cuculi

A relative of the stitchworts and campions, Ragged-robin can be found in wet meadows and fens and sometimes in damp spots in woodland. The five petals are deeply cut into distinctive fingered lobes and the flowers are carried in loose clusters on stems growing to around 70 cm in height. The paired leaves are narrow and finely pointed. Flowers are usually rich pink, but sometimes white.

Great Yellow-cress
Rorippa amphibia

A mass flowering of this plant around a flooded pingo in Breckland is one of the finest sights of May. It has the yellow, four-petalled flowers that are typical of its family, but differs from its relatives in its leaves, which are simple, without side leaflets, and variably sharply toothed along the margins. Underwater leaves may be very finely cut right to the veins. A tall species which may grow to a metre in height and which can be common along riverbanks and pond sides, usually growing from the water itself.

Mike Crewe

Marsh Yellow-cress
Rorippa palustris

Similar to Creeping Yellow-cress, but this is an annual plant that does not form creeping mats. The leaves are very variable, the lower ones often lance-shaped and toothed, while the upper are deeply divided into pinnate lobes. The bright yellow flowers are more showy than those of Creeping Yellow-cress, with the petals up to twice the length of the sepals. Found in a wide range of damp places, usually on bare mud beside streams, ditches and other waterways.

Mike Crewe

Creeping Yellow-cress
Rorippa sylvestris

A rather scarce plant in Breckland, but one which occasionally turns up on bare damp ground and may sometimes be found in disturbed conditions such as field margins and muddy woodland tracks. Plants grow to around 60 cm in height and when established may often spread to form creeping mats of vegetation. The leaves are pinnate, with strongly toothed leaflets, and the flowers are small, with petals only about the same length as the sepals.

Mike Crewe

Wetland

Common Water-cress
Nasturtium officinale

Although perhaps best-known as a cultivated plant, Water-cress also grows in streams and ditches as a native plant. A low, spreading plant, forming mats of pinnate leaves that grow out over the water surface from creeping stems. Clusters of white flowers appear just above the leaves and are followed by narrow seedpods that have seeds in a double row on each side.

Similar species: Narrow-fruited Water-cress (*N. microphyllum*) is very similar but much less common. The seeds in its pods are borne in a single row.

Mike Crewe

Wavy Bitter-cress
Cardamine flexuosa

A very common annual plant to 50 cm in height and found in all kinds of damp places, most often in shady spots and occasionally as an urban or garden weed. The leaves are pinnately lobed, mostly in a basal rosette with a few stem leaves. The flowers appear at the top of slightly wavy stems and are followed by long seedpods. The flowers have six stamens (hand lens required!).

Similar species: Hairy Bitter-cress (page 67) is very similar, but the flowers have only four stamens.

Mike Crewe

Large Bitter-cress
Cardamine amara

This is an uncommon plant in the Brecks and most likely to be found in damp shady places on acid soils along the Rivers Lark and Wissey. A perennial species, it can often form quite extensive colonies of bright green, pinnate leaves from creeping rootstocks. The flowers are showy, larger (up to 15 mm across) than those of other bitter-cresses, and with violet anthers at their centres.

Mike Crewe

21

Cuckooflower
Cardamine pratensis

Often known as Lady's-smock, this plant was once very common in wet meadows, grassy fens and along damp woodland rides. Now much less abundant, it can still be found on some of the area's commons and in river valleys. The leaves are pinnate, the upper ones much narrower than the lower. Flowers are relatively large – up to 18 mm across – and are usually pink, but sometimes white.

Square-stalked St John's-wort
Hypericum tetrapterum

A perennial plant that occurs in a wide range of wetland habitats, from marshes, fens and wet meadows to river and ditch banks. Stems grow to around 70 cm in height and bear simple, paired leaves. The stems themselves are strongly ridged and four-sided, being square in cross-section. The angles of the stems are slightly winged.

Similar species: See **Part One** for other St John's-worts, which have rounded stems.

Marsh Cinquefoil
Comarum palustre

Sadly, this interesting species of muddy fens and wet meadows has been lost from many of its former sites, but you still might come across it on some of the protected reserves along the Little Ouse valley and around pingos in the Wissey watershed. Growing to around 40 cm in height, the stems trail through very wet places, often on the edge of open water, and bear dull green leaves with five or seven leaflets. Flowers are a dark reddish-purple colour, with five petals.

Wetland

Water Dock
Rumex hydrolapathum

Docks are more leaf than flower, but this species still makes an impressive sight due to its large size. Plants may grow to two metres in height, with long, lance-shaped leaves up to a metre in length. The flowers are green and tiny, but they are carried in great quantity in branched spikes. Distinctive by its size alone, it may be found growing along the edges of permanent water bodies such as rivers, ditches and ponds.

Clustered Dock
Rumex conglomeratus

A slender, much-branched species of dock that grows commonly in all kinds of damp muddy places. It may often be found in ditches and on riverbanks as well as along damp field edges. Leaves are narrow with tapered bases, and stems grow to around 80 cm in height. The many small green flowers mature into red fruits, each having three swollen bumps.

Similar species: Compare with Wood Dock (page 50) and other docks (see Part One).

Golden Dock
Rumex maritimus

A distinctive species that grows on damp mud in the Lark valley and around seasonally wet pingos in the Norfolk Brecks. Plants are usually short, to 50 cm in height, and end in many-branched spikes, crowded with long leafy bracts and masses of green flowers. The flowers turn bright yellow-brown as they ripen and are followed by fruits with distinctive long teeth.

Similar species: Marsh Dock (*R. palustris*) is a scarce plant found in similar places. It usually grows taller and has shorter teeth on the fruits.

Water Chickweed
Myosoton aquaticum

A common species found in a wide range of wetland habitats, especially on peaty soils, and often on dredgings where ditches have been cleaned out. Usually a low, sprawling species, but stems sometimes scramble up through taller vegetation. Flowers are larger (to 15 cm across) than other members of the chickweed family. Leaves are pointed, wavy-edged and broad-based, with the bases clasping the stems in pairs. The five petals are divided almost to the base.

Marsh Stitchwort
Stellaria palustris

A rather rare species in Breckland, found in a few wet and well-preserved fens and marshes. This is a slender plant with square stems that may grow to 60 cm in height and scramble through other plants. The plant has a blue-green colour and the leaves are narrow and fine-pointed. Small leaf-like bracts beneath the flowers are pale with a central green stripe.

Similar species: Lesser Stitchwort (see Part One) often grows in damp grassland. It is greener, has narrower petals and no green stripe on the bracts.

Bog Stitchwort
Stellaria alsine

This species is rather local in its distribution, being more common in the Norfolk Brecks than on the Suffolk side, but it may be found in quite a few places in wet meadows and fens and also in woodland on damp tracksides. A smaller plant than other stitchworts, creeping low to the ground. The five petals are deeply divided, with the two halves of each petal forming a series of wide V-shapes. The smooth leaves and stems are blue-green in colour.

Wetland

Knotted Pearlwort
Sagina nodosa

Quite a widespread species on damp soils in heathy areas, especially around ponds and pingos that dry out periodically. A low, creeping plant with flowering stems rising to 10 cm or so in height. On the upper stems, the linear and slightly succulent leaves are carried in small clusters, looking like knots in a piece of string and giving the plant its name. Easily overlooked because of its small size.

Water-violet
Hottonia palustris

A beautiful plant that grows floating in standing water. Plants spread to form a mat of intertwined stems, the leaves all submerged and finely cut to the veins. In May, the plants send up vertical shoots, bearing whorls of pale lilac-pink flowers (pale enough to look white from a distance) which appear in great number and put on a spectacular show. A rather scarce plant due to land drainage schemes, but it may still be found in ponds and pingos in the Thet and Wissey watersheds.

Brookweed
Samolus valerandi

This little plant with its tiny flowers is easily overlooked. Rather rare in Breckland, it may be found on the muddy margins of ponds, ditches and streams, usually on acidic soils. Plants grow to around 30 cm in height from a basal rosette of broadly oval, pale green leaves. The tiny flowers, just 3–4 mm across, open a few at a time on elongated flowering spikes. They are followed by rounded seed capsules, the lower ones developing while the upper flowers are still open.

Bog Pimpernel
Anagallis tenella

A lovely but sadly rare species in our area, confined to a handful of protected peaty bogs in the Little Ouse valley and the Wissey watershed. It creeps low through mossy bogs, the thread-like stems carrying rounded leaves in pairs along their length. The stems produce large numbers of small pink flowers, each with five or six finely red-veined petals.

Fen Bedstraw
Galium uliginosum

Bedstraws have slender square-sided stems bearing whorls of leaves and scramble through surrounding vegetation. Fen Bedstraw can be told from similar species by its leaves, which have finely pointed tips and rows of backward-pointing bristles along the edges. The stems also have stiff bristles along their corners. A widespread plant of fens and wet places with chalky groundwater.

Marsh Bedstraw
Galium palustre

This species is very much like Fen Bedstraw at first glance, but is easily differentiated by a careful check of its leaves and stems. Marsh Bedstraw grows to around a metre in height and scrambles through neighbouring vegetation, but it may also be found growing as a low plant in shorter vegetation or on the edges of open water. The whorled leaves have smooth edges with no prickles and have blunt tips. The stems are smooth-sided without bristles. Usually found in areas with acidic groundwater.

Wetland

Water Forget-me-not
Myosotis scorpioides

Found along river and stream banks, pond-sides and in very wet fens. A relatively large wetland forget-me-not which is perennial and may grow to 60 cm or so in height, but which may also be found trailing from the bank into open water. The flowers appear in flat-topped clusters, each flower being around 8–12 mm across. The calyx (the green part at the base of the flower, below the petals) has relatively short triangular teeth, which are as broad at the base as they are long.

Mike Crewe

Tufted Forget-me-not
Myosotis laxa

An annual or biennial species that is much smaller and less common than Water Forget-me-not and grows in unimproved wet meadows and fens. Stems grow to around 40 cm in height (or less) and carry an open cluster of small flowers at the top. The flowers are only around 4 mm across. The calyx (the green part at the base of the flower, below the petals) has relatively long triangular teeth, which are clearly shorter at the base than they are long.

Mike Crewe

Brooklime
Veronica beccabunga

The brilliant blue speedwell flowers of Brooklime are very eye-catching and this plant can be found in a wide range of wetland habitats, wherever there are bare mud patches. It can also occur as an arable weed if the ground is wet enough in low hollows and gateways. The flowers appear in spikes from the leaf and axils, and the stems trail on the ground, rooting as they go. The broadly rounded leaves and the stems are shiny and hairless.

Mike Crewe

Blue Water Speedwell
Veronica anagallis-aquatica

A common species found growing beside ponds, streams, rivers and ditches. Plants grow to a height of around 60 cm and the lower leaves are broad and have short petioles (leaf stalks), while the upper leaves are lance-shaped and without petioles. The pale blue flowers are carried in upright spikes from the upper leaf axils. Each flower has a leafy bract at its base which is shorter than the individual flower's stalk when the flower is fully open.

Pink Water Speedwell
Veronica catenata

A common species which grows in similar places to Blue Water Speedwell, but which is less often found by faster-moving water. Plants grow to a height of around 60 cm and all of the leaves are lance-shaped and without petioles. The pink flowers are carried in upright spikes from the upper leaf axils. Each flower has a leafy bract at its base which is longer than the individual flower's stalk when the flower is fully open.

Marsh Speedwell
Veronica scutellata

This species is much less common than the two water speedwells and is found in undisturbed wetland habitats where the soils are a little more acidic. A variable species, but most often found with the stems trailing rather than upright and reaching 50 cm or so in length. Leaves are lance-shaped, in opposite pairs. The flowers appear in long spikes like those of the water speedwells, but they are carried alternately from the leaf axils and not in opposite pairs. Individual flowers are pale, almost white, with purplish veins.

Wetland

Water Figwort
Scrophularia auriculata

A tall plant, growing to a metre in height and common in many kinds of wetland habitats, including roadside ditches. The leaves are heavily crinkled and have rounded bases, often with an extra pair of small leaflets on the leaf stalk. The main stems are four-cornered and narrowly winged. Flowers are tiny and carried in a many-branched head at the top of the plant.

Similar species: Green Figwort (*S.umbrosa*) is almost identical but rarer. Its stems are more strongly winged and the leaves more pointed at the tip and have sharply toothed margins.

Bittersweet
Solanum dulcamara

An attractive member of the nightshade family which has richly purple flowers with a yellow centre, followed by bunches of bright berries that turn from green through yellow to red. The leaves are variable and may be simple or have basal lobes. Also known as Woody Nightshade, this is a plant of many kinds of wetland habitats, most commonly found on the edges of reedbeds but also as a weed of damp and disturbed places, even in towns and gardens.

Marsh Pennywort
Hydrocotyle vulgaris

This little plant of wet fens and bogs is unlikely to be noticed by anyone looking for showiness, since its flowers are tiny (about 1 mm across), green and appear in small clusters beneath the leaves. However, the circular leaves (up to 5 cm across) grow from creeping stems that root at the nodes and form quite extensive mats of vegetation beneath other plants – their presence is common enough to attract enquiring minds. Have a poke around and see if you can find the flowers!

Water Mint
Mentha aquatica

A common plant of damp places and often forming large colonies. It can often be detected long before it is seen, from the familiar mint smell that wafts up from underfoot – give a leaf a rub and a sniff! Square-sided, upright stems grow to around 90 cm in height and carry opposite pairs of very hairy leaves. Tight bundles of pale lilac-pink flowers are clustered in the leaf axils. Stems are topped by a cluster of flowers.

Similar species: Corn Mint (page 83) has stems topped with small leaves.

Gypsywort
Lycopus europaeus

A common plant of damp places, especially in tall fen vegetation and along river and ditch banks. The square-sided stems and bristly hairy leaves are typical of the dead-nettle family, but the leaves are distinctive in being sharply toothed along the margins. The individual white flowers are tiny, but carried in tight bundles in the leaf axils towards the top of the stems. Plants may branch and become quite bushy, growing to nearly a metre in height.

Marsh Lousewort
Pedicularis palustris

A scarce plant of very wet acidic bogs, in Breckland being more or less confined to protected reserves in the upper Little Ouse valley and along the Wissey. This is a fairly small semi-parasitic plant, its branched stems growing only to 25 cm or so in height and often much less. The flowers are pinkish-purple with a tubular base, which opens out into a two-lipped mouth, somewhat like a small snapdragon. Its stems carry many finely cut, fern-like leaves.

Wetland

Common Skullcap
Scutellaria galericulata

This attractive little plant is rather rare in Breckland's wetland areas and may easily be overlooked in taller vegetation. The delicate stems with narrow, blunt-toothed, opposite leaves grow to around 40 cm in height and the flowers appear in pairs towards the top of the stems. The flowers are deep violet-blue and have a tubular base with a two-lipped mouth.

Similar species: Compare with Ground-ivy and Self-heal (see Part One), both of which typically grow in drier habitats.

Marsh Woundwort
Stachys palustris

A fairly common wetland plant along river corridors of the Brecks, becoming more common westward into fenland habitats. Closely related to the dead-nettles and rather like them, it produces stems to a metre in height, bearing narrow, stalkless, lance-shaped leaves in opposite pairs. The tubular flowers are deep pink, with the outer lips bearing an intricate pattern of pale marks.

Similar species: Hedge Woundwort (see Part One) has darker flowers and its broader leaves have rounded bases on obvious leaf stalks.

Devil's-bit Scabious
Succisa pratensis

This is a rather late-flowering species that can brighten up a Breckland walk after many other flowers are past their best. A species of unimproved wet meadows and acid flushes, the stems grow to 80–90 cm in height with the flowers appearing in tightly packed clusters at the top. The lightly hairy leaves are narrowly oval with tapered bases.

Similar species: Other scabiouses (see Part One) grow in drier habitats and have pinnately lobed leaves.

Meadow Thistle
Cirsium dissectum

A rare species in the Brecks, which avoids chalky groundwater and is largely confined to the more acidic wetlands in the Little Ouse valley and in the Foulden area. It is much more gentle than other thistles, having spineless, white-downy stems to 60 cm in height that carry a single flowerhead at the top. The leaves are ovate with prickly margins and do not have the deeply divided look of most other thistle species.

Marsh Thistle
Cirsium palustre

A very common species of wetland habitats, including damp woodland. Plants grow from very spiny basal leaf rosettes, with stems sometimes reaching two metres in height, although often less. Flowers appear in tight clusters at the stem tops, their outer spiny bracts being flattened against the head. Stems have spiny wings and the whole plant is often purple-tinged.

Similar species: Welted Thistle (see Part One) is less common and has similarly winged stems. The outer bracts of its flowerheads are bristly.

Butterbur
Petasites hybridus

Always a surprising plant to find, as the flower spikes emerge direct from the ground in spring and may reach 40 cm in height. As the flowers begin to die down, the leaves start to emerge, gradually expanding to become nearly a metre across and rather resembling rhubarb leaves. Generally uncommon, but it can be locally plentiful along stretches of the Little Ouse and the Black Bourn.

Similar species: Winter Heliotrope (see Part One) has the same growth style, but smaller leaves and smaller, more open flower spikes.

Wetland

Nodding Bur-marigold
Bidens cernua

A once widespread species that is now rare in Breckland and most likely to be found in parts of the upper Wissey watershed. This is a plant of muddy water margins, growing particularly in bare areas where winter floodwater draws down in summer. Branching stems carry simple lanceolate, paired leaves with coarsely serrated margins, with stems growing to around 80cm in height. Nodding flowerheads are usually without an outer array of yellow petals, but these may be present occasionally.

Trifid Bur-marigold
Bidens tripartita

As with Nodding Bur-marigold, this species has declined greatly in recent years and needs careful searching to find in the upper reaches of the Wissey watershed. A plant of muddy margins, especially on winter-wet land that dries out in summer. The leaves are mostly divided into three lobes, the middle lobe being the largest. The flowerheads are similar to those of Nodding Bur-marigold, but remain more or less erect and not nodding.

Marsh Ragwort
Senecio aquaticus

A rare species in Suffolk Breckland but a little more frequent in Norfolk in the Wissey watershed. An upright species, growing to around 80 cm in height. Lower leaves are pinnately lobed, like other ragwort species, but Marsh Ragwort has a broad, rounded end lobe to its leaves, which is clearly larger than the side lobes. The yellow flowerheads are borne on an open, broadly branching flower spike.

Similar species: Common Ragwort (see Part One) has leaves with lobes more similar in size and more flowers, in a denser flowering stem.

Hemp-agrimony
Eupatorium cannabinum

A common and very showy plant found in all kinds of damp habitats including roadside ditches and wet field margins. Tall, growing to 150 cm in height and forming a stout clump. The stiff stems carry coarsely toothed leaves that are usually three- to five-fingered. The tiny pink flowers appear in great numbers in large, flat-topped 'mop heads', which are very attractive to butterflies and other insects.

Common Valerian
Valeriana officinalis

A fairly common species which occurs both in tall wetland plant communities and occasionally in drier grassy places. The stems may grow to over a metre in height with leaves that are pinnately lobed, strongly toothed along the margins and with prominently indented veins. The tiny five-petalled flowers (5 mm across) have tubular bases and are carried in great number above the foliage in dense, rounded clusters. The sweet-scented flowers are pink in bud, opening to more or less white in colour.

Marsh Valerian
Valeriana dioica

A less frequent species than Common Valerian, but locally common in unimproved wetland meadows and fens. A small plant, usually to just 30 cm in height, and with stems arising from a basal cluster of simple spoon-shaped leaves. The stem leaves differ from these in being pinnately lobed. The tiny flowers, just 5 mm across, are deep pink in bud, opening to pale pink.

Wetland

Hemlock
Conium maculatum

A very common species typically found in damp nutrient-rich places along roadsides, dredged ditches and channels and wet field margins. A tall and imposing plant, which may grow to over two metres in height and has a strong and unpleasant smell like that of male mice. The leaves are very finely cut into lacy segments and the stems have deep purple blotches on them. **A poisonous plant – wash hands after touching!**

Similar species: See Cow Parsley and relatives in **Part One**.

Fool's Water-cress
Apium nodiflorum

Common in suitable habitat throughout the Brecks in ditches, river edges and streams. Plants grow rapidly from the bank into open water and can often clog up smaller streams and ditches, with the stems rooting from the leaf nodes as they sprawl. Easily confused with Water-cress (page 21) when not in flower due to its similar leaves, and because it often grows with it. Tiny white flowers are carried in small umbrella-like clusters along the stems, with each cluster appearing opposite a leaf.

Lesser Water-parsnip
Berula erecta

Widespread in wet ditches, streams and river margins but less common than Fool's Water-cress and usually not in such abundance. Very similar to that species, but usually more upright and with leaves having more sharply toothed margins. A careful check of the basal stem of the leaf should reveal a pale ring which is absent in Fool's Water-cress. Tiny white flowers are carried in rounded clusters opposite a leaf.

Wild Angelica
Angelica sylvestris

A stocky plant of damp habitats, including fens, wet meadows, wet woods and shady roadsides. It has a stout, prominently ribbed stem that may grow to two metres in height. The leaves are only coarsely divided into ovate leaflets and the leaf stalks are prominently inflated and sheath around the stems. The flowers are carried in rounded heads, becoming flat-topped with age, and each cluster emerges from a distinctive sheath-like bract.

Giant Hogweed
Heracleum mantegazzianum

A massive and imposing plant that can reach four metres in height! It has spread from cultivation into the wider countryside and although still uncommon in Breckland, is slowly increasing, especially along the Lark around Mildenhall. Huge, coarsely toothed leaves grow to over two metres in length and are topped by stout, strongly ribbed stems that have purple blotches on them. Flowers appear in large, flat-topped, many-branched heads. The sap can cause dermatitis and open sores when in contact with skin.

Milk-parsley
Thyselium palustre

Mostly a plant of Norfolk's Broadland, this species can still be found in a handful of well-maintained Breckland fens in the upper Wissey watershed around the Great Hockham area. An upright hairless species, with smooth green stems growing to around 150 cm in height. The leaves are several times divided into smooth-edged leaflets and the flowers are carried in flat-topped clusters at the top of the plant.

Wetland

Tubular Water-dropwort
Oenanthe fistulosa

A rather small and delicate species that can be missed easily among taller vegetation. Rather scarce in the Brecks, but may be found on the margins of permanent water, especially around some of the ponds and pingos. A shiny and smooth plant, the stems growing to around 60 cm in height. The long stalks of the leaves are hollow, while the leaves themselves are divided almost to the veins into very narrow leaflets. Flowers are carried in relatively small branched heads, usually with only three or four branches to each head.

River Water-dropwort
Oenanthe fluviatilis

A difficult plant to find at times as it grows as a submerged aquatic plant in permanent rivers and larger streams until it reaches flowering size, when it appears above the water surface. Most common in the Little Ouse around Thetford. Could be mistaken for Fine-leaved Water-dropwort, but it has broader leaf segments rather like flat parsley. The flat-topped flowerheads are at first held upright at the top of the stems, but fluctuating water levels can turn the plant into a tangle of stems, leaves and flowers.

Fine-leaved Water-dropwort
Oenanthe aquatica

A very local species in Breckland, being absent from many seemingly suitable locations, but plentiful in the pingos around Thompson Common, where it can put on a great show in late June and July. Like River Water-dropwort, this plant starts as a submerged aquatic but it prefers ponds to moving water and sometimes occurs in pools that dry out in summer. Submerged leaves are deeply cut right to the veins, while upper leaves are deeply cut and fern-like.

Military Orchid

Orchis militaris

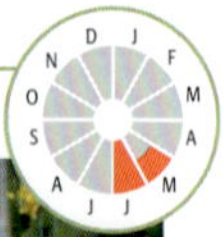

Mike Crewe

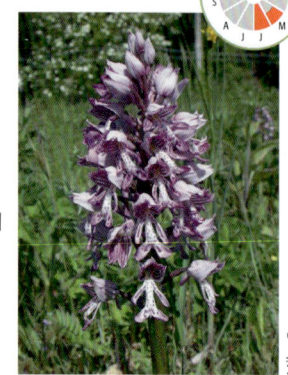

A book on the flowers of Breckland would hardly be complete without including this magnificent orchid – but it is not a plant that you are likely to chance across during a casual woodland stroll. The discovery of the Military Orchid in an old chalk pit near Mildenhall in 1954 was one of the great botanical revelations of its day and this special place is now managed by the Suffolk Wildlife Trust as a reserve for this and other scarce plants of the area.

Military Orchid flowers get their name from a rather fanciful impression that they look like little people wearing an old type of military helmet. The flower spikes may grow to about 30 cm in height and look quite spectacular in full flower. The leaves are plain green, without spots or dark marks.

Details of the special open day (usually in late May) to allow visitors to see this plant can be obtained from the Suffolk Biodiversity Information Service website, or by contacting the Forestry Commission.

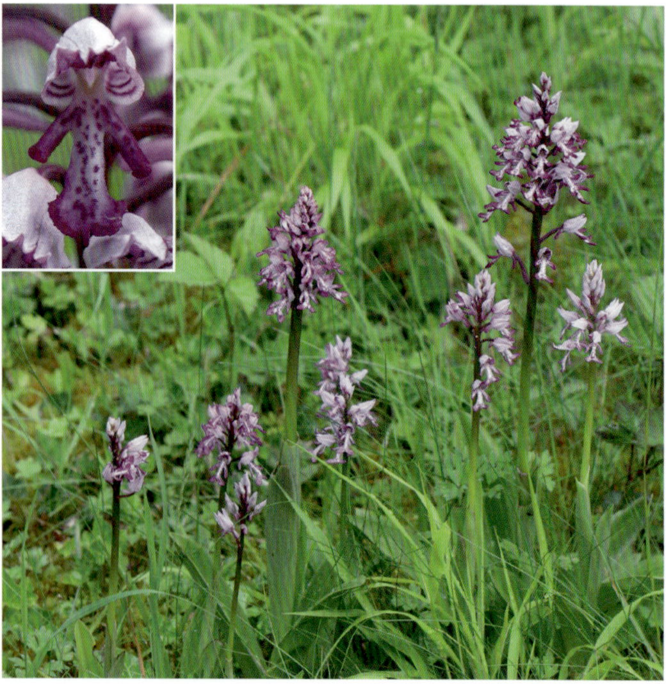

Main image, Mike Toms / BTO; Flower, Mike Crewe

Woodland

Yellow Star-of-Bethlehem
Gagea lutea

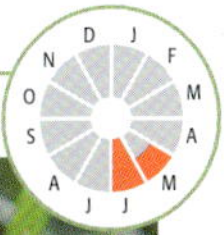

This very special plant can be found at a number of locations scattered throughout the UK, but only at a handful of sites in East Anglia. Only one of these is in Norfolk – Wayland Wood, just on the edge of the Brecks and where this species can be seen relatively easily. The plants can be found in the south-east corner of the site, towards the road.

Yellow Star-of-Bethlehem grows from a summer-dormant bulb, the plants spreading to form quite dense colonies of bright green, grass-like leaves, very similar to those of our native Bluebell. Plants can be a little shy to flower and usually there are many more plants with leaves than there are plants with flowers. The bright yellow flowers have a distinctive green stripe on the back of each petal and may either be solitary or appear in bunches of up to seven or so.

Broad-leaved Helleborine
Epipactis helleborine

This is generally a rather rare species in East Anglia, but several colonies are known to occur in deciduous woodland belts in forestry around Santon Downham. Flowering spikes can vary in height, from 30 cm to almost a metre, and the flowers are rather variable in colour, although they usually have at least some pink or reddish tints to them. The lower leaves are rather broad for an orchid and clasp the stem at their bases.

Similar species: Green-flowered Helleborine is very rare and differs in its smaller, nodding, all-green flowers.

Common Spotted Orchid
Dactylorhiza fuchsii

Although a common orchid nationally, Breckland's sandy heartlands are too acidic for this species, so it is more likely to be found in damp grassland and deciduous woodland in the northwest and northeast of the area. The leaves are heavily spotted with dark reddish-purple and the flowers are white or very pale pink with purple spots. Note the rather narrow, deeply three-lobed lower lip of the flowers.

Similar species: See other spotted orchids on pages 12–13.

Early Purple Orchid
Orchis mascula

This species does best on the clay soils of East Anglia's middle section, but it may be found in the deciduous woods around the edges of the Brecks, especially south of Watton. A relatively early flowering orchid, with the flowers at their best in April. Leaves may be plain green, but they are more often blotched or spotted with dark purple. Flowers have a neatly folded lower lip and vary in colour from deep pinkish-red through pink to white.

Woodland

Common Twayblade
Neottia ovata

A species found in a wide variety of habitats from acid wetlands to chalky grasslands, but in the Brecks most likely to occur in woodland. Common Twayblade is very easy to miss, as the narrow spike (to 60 cm in height but often much less) of greenish-yellow flowers can be hard to spot amongst other foliage. More obvious are the broadly rounded leaves, of which there is just a single pair at the base of each plant.

Mike Crewe

Lords-and-ladies
Arum maculatum

The wealth of common names (Cuckoo Pint, Wild Arum, Adder's Root, and more) given to this plant indicates that it is a well-known and common species. The bright green, arrowhead-shaped leaves appear in mid- or late winter and often form quite large colonies. The unusual flowers, growing to 30 cm in height, are very distinctive and unlikely to be mistaken for any other native species. The flowers wither in late spring and are followed in the summer by a spike of bright red berries, sometimes known as dead man's fingers!

Mike Crewe

Dog's Mercury
Mercurialis perennis

An abundant species in deciduous woodland, where creeping colonies can be very extensive. In Breckland, suitable habitat is largely confined to the outer edges of the area. Stems grow to around 40 cm in height and spread by underground shoots to produce mats of leafy growth that colonise the woodland floor. In spring, emerging shoots carry small green and inconspicuous flowers in short spikes, the male and female flowers appearing on separate plants – look carefully for female plants, which are much less common than males!

Mike Crewe

Common Bluebell
Hyacinthoides non-scripta

Although in the UK Bluebells can be abundant in woodland on acid soils, the chalky soils and lack of ancient woodland in Breckland make this a rather scarce plant here, except in woods at the edges of the area. While our Common Bluebell is considered to be easy to recognise, care needs to be taken to distinguish it from hybrid plants that have spread from gardens (see below). Look for flowers that are pendulous, have strongly back-curled petal tips, and which grow in a one-sided, arching spike. Flowers are occasionally white or pink.

Mike Crewe

Hybrid Bluebell
Hyacinthoides x massartiana

Very popular as a garden plant and often wrongly labelled as 'Spanish Bluebells', these flowers are commonly found in disturbed areas, on roadsides or near houses. They are very variable plants, being hybrids between our native Bluebell and the Spanish Bluebell (see below), and flowers may be various shades of blue, or commonly pink or white. Look for plants with flowers that are only slightly pendulous and arranged all around the straight (not arching) stem.

Mike Crewe

Spanish Bluebell
Hyacinthoides hispanica

This plant is rare in the Brecks, but occasionally occurs on roadsides and waste places where planted or allowed to spread from nearby gardens. Although rare, it is included here to help with correct identification of the various bluebells in the area. It has much broader leaves than those of our native Bluebell, and upright spikes that carry flowers on all sides. The flowers are broad with petals that are almost straight (not curved backward) and usually pale blue.

Mike Crewe

Woodland

Greater Celandine
Chelidonium majus

As an ancient introduction, this is not so much a woodland plant but one of shady places around human habitation – it may often be found growing from old walls, on shady banks and even pavement cracks. Plants reach around 75 cm in height and can easily be told by the bright yellow-orange sap that appears if a piece of one of the compound leaves is broken off. The yellow flowers are followed by short, green seedpods.

Lesser Celandine
Ficaria verna

A species that avoids the drier sandy conditions of the Breckland heaths but can be found in damper woodland soils and along banks and lanes in the river valleys. The brilliant yellow flowers usually have eight petals, but any number up to 12 is not uncommon. It can form often quite extensive mats of colour when in flower and is one of the earliest spring flowers to appear in the area. Leaves may be green, or variously marked with dark or pale patches.

Winter Aconite
Eranthis hyemalis

This is not one of our native species, but so eye-catching when it comes into flower in late winter that it is bound to be noticed by early season walkers. Aconites grow from a tuber, with each one producing a single flower accompanied by a green 'ruff' of leafy bracts. Plants can spread to form extensive colonies in a riot of colour beneath trees in shady churchyards and wooded spots near houses. Often found with Snowdrops and flowers around the same time.

Enchanter's-nightshade
Circaea lutetiana

This low plant of shady places occurs in woods and along hedge bottoms where woodland once stood. The delicate spikes of tiny pinkish-white flowers grow typically to around 30 cm in height, each flower bearing just two deeply notched petals. The flowering stems are covered in slightly sticky glandular hairs and the flowers are followed by rounded seedpods that are covered in hooked hairs, designed to hitch a ride on passing animals – or unsuspecting socks!

Small Balsam
Impatiens parviflora

An uncommon introduction that may be found in shady disturbed places in a handful of locations, being most frequent on the eastern edges of the Brecks around Harling and Hockham. Plants may grow to almost a metre in height, but are often much shorter. The stems and simple leaves are bright green, hairless and slightly succulent. The flowers are rather small, pale yellow and have a long spur containing nectar at the rear. The slender pods explode suddenly if touched, throwing their seeds far and wide.

Moschatel
Adoxa moschatellina

This tiny unassuming plant creeps across the woodland floor to form leafy patches up to 15 cm in height. The leaves are divided into three deeply lobed segments. Although this is one of the earliest wildflowers to appear in spring, its diminutive size and green colouration mean that it is easily missed without careful searching. The old country name of Townhall Clock alludes to the arrangement of four flowers facing outward, like the faces on a clock tower.

Woodland

Wood Anemone
Anemone nemorosa

A plant of loamy woodland soils, rare in the Brecks, but its flowers can be found trembling in the spring breeze in woods out towards Watton and Swaffham. A low-growing plant that forms carpets of leaves across the ground and has flowers with six or seven petal-like sepals that are white on top and pink-washed beneath. The leaves are three-lobed and deeply cut.

Wild Strawberry
Fragaria vesca

There can be few people who have not been strawberry picking and this little plant of shady places and woodland paths is just like a small version of the familiar garden strawberry. The small white flowers have slightly pointed petals and are carried in loose bunches above the leaves. These are followed by tiny strawberry fruits, no more than 1 cm or so in length – but they are still tasty! The three-lobed leaves are hairy and strongly veined.

Wood-sorrel
Oxalis acetosella

This species prefers acidic soils and tends to avoid the areas of central Breckland that are influenced by the chalky bedrock. The distinctive pale-green, three-lobed leaves are considered to be one of the possibilities for the true identity of the Irish Shamrock. The cupped flowers rise above the leaves on slender stems and the white petals are neatly veined in pale lilac.

Wood Avens
Geum urbanum

A common and widespread species of woodland and other shady places. The leaves are very variable, the lower ones being pinnate and having large, rounded end lobes, while the upper stem leaves are simple or three-fingered. Stems grow to 60 cm or so in height and carry small yellow flowers whose petals seem undersized, with the green sepals showing between them. The flowers are followed by a reddish spiky seedhead, bearing hooked bristles.

Water Avens
Geum rivale

This species is closely related to Wood Avens and the leaves look very similar, being rounded and pinnate on the lower stem and simple or three-fingered on the upper part of the plant. However, the flowers are very different, with those of Water Avens being peach-coloured and carried on graceful, nodding stems. A plant of wet shady places, often growing in muddy patches in woods or beside shaded streams.

Hybrid Avens
Geum x intermedium

While Water Avens and Wood Avens look very different to each other, they are closely related and, where they grow in close proximity, they will often hybridise. The resulting hybrids are very variable as they try to be something halfway between their parents! Flowers may be yellowish or pinkish and may be nodding, upright, or somewhere in between. Look for them in places where you find Water Avens in wooded spots.

Woodland

Common Dog-violet
Viola riviniana

The commonest woodland violet in Breckland (although beware of Sweet Violet, which commonly spreads from gardens), growing in a wide range of shady places. The flowers show relatively broad petals when viewed from the front, while a side view reveals long, pointed sepals and a pale whitish spur at the back of the flower. The leaves are bright green and more or less hairless.

Early Dog-violet
Viola reichenbachiana

A widespread species in the Brecks, but less common than the above species. Viewed from the front, the petals are relatively narrow and scarcely overlap each other, while a side view reveals pointed sepals and a deep purple spur. Leaves are bright green and more or less hairless.

Similar species: Compare both the woodland dog-violets with violets of grassy places (see Part One).

Climbing Corydalis
Ceratocapnos claviculata

This plant is typically found on acidic heaths with plenty of Bracken and will grow in shade after the heaths have been planted with conifers. Most common on the heaths to the northeast of Thetford, it is a scrambling plant, often growing to a metre or more in height as it trails up through other vegetation. Often tight heads of creamy-white tubular flowers can be found throughout much of the year and appear above delicate fern-like foliage.

Herb-Robert
Geranium robertianum

A widespread and common plant, found in all kinds of shady places, sometimes even growing out of walls and cracked paving. The bright pink flowers appear on long stems which may grow to 30 cm or more in height. The lower stems are much-branched and carry leaves that are typically made up of five deeply and finely cut leaflets. Being a relative of the crane's-bills, the flowers are followed by long, pointed seedpods.

Mike Crewe

Wood Forget-me-not
Myosotis sylvatica

This plant may occasionally be found in deciduous woodland on loamier soils, but it is more likely to occur in shady places around churchyards, grassy banks and similar places, as this is the forget-me-not that is so popular as a garden plant. The bright, pale blue flowers appear in clusters above the leaves – but may be white or pink in some garden forms. The stems and simple leaves are covered in bristly hairs.

Mike Crewe

Wood Speedwell
Veronica montana

A low, creeping plant of shady woodlands and tracksides. The pale bluish flowers have distinct purplish veins on them and appear in small spikes in the upper leaf axils, the flowers opening one or two at a time. The leaves are deeply toothed around the margins and are pale yellowish-green in colour, while the stems have whitish hairs evenly spread all around.

Similar species: Compare stem hairs and flowers with Germander Speedwell (see **Part One**).

Mike Crewe

Woodland

Three-nerved Sandwort
Moehringia trinervia

A low-growing plant with slender stems that trail close to the ground or scramble through other vegetation. The chickweed-like flowers have five rounded, white petals that are shorter than the long-pointed green sepals. The oval leaves have three well-defined veins which usually can be best detected from the underside of the leaf.

Similar species: Chickweeds and stitchworts have a single well-defined vein in each leaf.

Berry Catchfly
Silene baccifera

This plant was introduced to the Brecks many years ago and may still be found in woodland and along shady roadsides around the village of Merton. The sprawling, branched stems with simple leaves scramble through other vegetation and may grow to a metre in length. The unusual flowers have swollen bases and five deeply notched petals. After flowering, the seeds develop in a green berry, which eventually turns black.

Red Campion
Silene dioica

Although mostly absent from Breckland's sandier heartlands, this species is common in woodlands and shady places on loamier soils along the river valleys and in the area's outer limits. Stems grow to around 90 cm in height, carrying broadly oval, deeply veined leaves. The reddish-pink flowers have five deeply notched petals and appear in small clusters at the tops of the stems.

Lily-of-the-valley
Convallaria majalis

This familiar plant of gardens is rather rare as a native species, but quite well established in one or two shady places in Breckland. It spreads by underground shoots to form creeping patches that send up dense stands of broadly oval leaves in the spring. The leaves are accompanied by flowering stems that grow to around 30 cm in height, the flowers being brilliant white, bell-shaped and famously fragrant. They may be followed by one or two bright red berries.

Black Bryony
Tamus communis

As a plant that favours chalky soils, this species is absent from the heathlands of Breckland and most likely to be found in shady places along the northern and eastern edges of the area. It is a climbing plant that grows up from the ground afresh each year. The stems have no tendrils, but climb by twining around neighbouring vegetation. The heart-shaped leaves are very glossy and the tiny flowers are carried in small clusters along the stems. Female plants bear shiny red berries in the autumn.

Wood Dock
Rumex sanguineus

Docks are common plants in the countryside and their large spear-shaped leaves are well known by most people. This is a rather delicate species, growing to around 60 cm in height in shady places. It is best told from other docks by the fruits that develop after the flowers; these have smooth edges without pointed lobes and each has a single red tubercle.

Similar species: Compare the fruits with those of other docks on page 23 and see Part One.

Woodland

Primrose
Primula vulgaris

The pale yellow flowers of Primroses are one of spring's great delights and they can be found in the richer moister soils of Breckland's northern edges. Popular as a garden plant, the species may also be found in churchyards and hedge banks near houses, and garden forms may sometimes be pink or white with a yellow centre. True primroses differ from the garden 'Polyanthus' types by having each flower arising singly from the base of the plant.

Yellow Pimpernel
Lysimachia nemorum

Growing in richer, loamy soils in damp woodland, this species is rather scarce in Breckland, and is most frequent in the deciduous woods around the perimeter of the region. The bright, cheerful flowers of Yellow Pimpernel grow on low stems that trail across the ground. The flowers are carried singly in the axils of the leaves, while the leaves appear in opposite pairs along the stems. The leaves are narrower than those of Creeping-Jenny and the flowers are longer-stalked, more open and star-shaped.

Creeping-Jenny
Lysimachia nummularia

This species is a native of wet shady places and is most likely to be found in damp soil along the watercourse of the Wissey. Its trailing stems creep across the ground and carry broadly rounded leaves with wavy edges. The five-petalled flowers are cup-shaped and appear singly from the leaf axils. Popular as a garden plant, it may also be found around churchyards and near houses where it has spread from cultivation.

Bugle
Ajuga reptans

A plant of shady places, this species forms low, creeping colonies along woodland tracks and the banks of quiet streams. The creeping stems have shiny, crinkled leaves in pairs, the leaves often having a purple tinge to them. The flowers appear in tight leafy whorls on upright stems, growing to around 30 cm or less in height. Flowers are bluish-violet in colour, but may occasionally be pink or white.

Yellow Archangel
Lamiastrum galeobdolon

The Yellow Archangel grows in shady places on open woodland floors. A relative of the dead-nettles, it has rich green nettle-like leaves and can form quite extensive patches. In spring, upright shoots carry bright yellow flowers with orange spots on whorls in the leaf axils.

Similar species: Compare with the garden form of grassy places, which has variegated leaves (see **Part One**).

Yellow Figwort
Scrophularia vernalis

This introduced species is very localised in Breckland, with small populations being well established in several places along the Lark Valley and in a few spots in the Hockham area. The stems may grow to a metre in height and could easily be mistaken for a sturdy dead-nettle, with their broad, well-toothed leaves. The small flowers are peculiar little things that look like inflated bladders and are carried in small bunches towards the top of the plant.

Woodland

Betony
Betonica officinalis

Sadly this species has declined greatly and is far less common in East Anglia than it once was. Keep an eye out for it in shady places on the loamier edges of the Brecks. Plants start with low, creeping stems that form patches of dark green, with taller, stiff flower stems emerging and carrying deeply toothed leaves. The rich cerise-pink flowers appear in whorled clusters at the top of the upright stems, which grow to around 60 cm in height.

Wall Lettuce
Mycelis muralis

Although Wall Lettuce is sometimes found on old boundary banks in woodland, it is more likely to be found in other shady places, especially churchyards and old flint walls. Plants vary in height (according to growing conditions), sometimes reaching 80 cm or so, but often much less. The lower leaves are very sharply lobed and angular-looking. Small flowers appear in many-branched, spreading heads.

Similar species: Compare with Nipplewort (see **Part One**) which has hairier and more rounded leaves.

Sanicle
Sanicula europaea

A plant of old, largely undisturbed, deciduous woodland. It might be difficult to recognise Sanicle as a relative of Wild Carrot or Cow Parsley, but it does have a similar growth style. The leaves are broadly rounded and lobed, rather like those of the Crane's-bills, and are found mostly at the base of the plant. Flower spikes are branched towards the top, with the tiny white flowers appearing in very shortly branched heads. Flowers are followed by small fruits that bear hooked bristles on their sides.

Star species

Smooth Rupturewort
Herniaria glabra

This is an odd little plant, which seems to do best on disturbed and compacted soil on gravel tracks, in pavement cracks and on old excavations, but sometimes it will turn up on the edges of bare arable land or as a garden weed. It is occasionally found scattered throughout much of the British Isles, but as a native species is considered to be confined to Breckland and an outlying area of Lincolnshire. It is classified as a Red Data Book species on account of its rarity.

Small and unobtrusive, it hugs the ground as it grows and its pale green stems with small, rounded, simple leaves may easily be overlooked. Even when in bloom it does little to attract attention, as the flowers are tiny – being just 2 mm across – and pale green in colour, scarcely standing out from the rest of the plant. They appear in tight clusters in the axils of the opposite, paired leaves.

Arable

Red-tipped Cudweed
Filago lutescens

This rare species seems never to have been common in the Brecks, but has proved to be persistent and continues to survive at a handful of locations in Suffolk. It will grow on sandy disturbed soil on field margins, as well as on dirt roads and on heathland where rabbit diggings provide the required disturbance. Exact places where it is best to look for this plant seem to change periodically, but field edges and disturbed tracks around Mildenhall and Elveden have most recently been the best sites for it.

As with other cudweeds, plants germinate in the autumn, producing a rosette of grey downy leaves during winter. The following year, an upright stem grows to around 20–30 cm in height, bearing elongated leaves with dense white hairs. At the top of the stems, the flowers appear in tight, ball-like clusters. Telling this plant from the far more widespread Common Cudweed can be extremely difficult and is best done by looking for the deep red hair-like tips to the outer bracts around the flowers.

Similar species: Compare very carefully with Common Cudweed (see **Part One**).

Star species

Breckland Speedwell
Veronica praecox

Speedwells are well known to many people as their cheerful rich blue flowers brighten up many a garden vegetable patch, flowerbed, field corner or churchyard. Breckland is blessed with an extra set of speedwells to those found elsewhere, although they have become rather rare and are most likely to be found at sites where special management is carried out to provide the right conditions.

This plant is a Red Data Book species because of its rarity. It still occurs at a handful of managed locations, but with careful searching may be found elsewhere on disturbed ground. The stems grow to around 15 cm in height and carry oval leaves with toothed margins. The leaves often look stiffer and darker green than those of other speedwells. The small flowers are deep blue, with all the petals being the same colour and carried away from the leaves on relatively long stalks.

Similar species: Compare very carefully with other speedwells of open ground on pages 82–83 and in **Part One**.

Arable

Spring Speedwell
Veronica verna

This species is one of a trio of rare speedwells of disturbed ground in the Brecks and largely owes its continuing survival to the work of the county wildlife trusts. Breckland is its only location in the UK, and even here it seems always to have been rare. A Red Data Book species that is classified as Endangered, it seems to survive in Norfolk only at Weeting Heath, where it has been sown and managed, while in Suffolk it still occurs along field edges and on disturbed areas on heaths, mostly at Mildenhall and Lackford.

A variable plant that may reach 15 cm in height, it is more often found growing to just 3–4 cm and bearing just one or two flowers. It greatly resembles the very common Wall Speedwell in having tiny flowers on short stalks that are well embedded among narrow leafy bracts, but differs in its lower leaves, which are deeply cut into three lobes. Despite its name, Spring Speedwell is typically the latest of the three, rare, open-ground speedwells of Breckland to flower in late spring or early summer.

Similar species: Compare very carefully with other speedwells of open ground on pages 82–83 and in **Part One**.

Star species

Fingered Speedwell
Veronica triphyllos

As plants of open, arable farm edges and other disturbed areas, the speedwells of the Brecks have declined greatly due to changes in farming practices and even the loss of farmland to new housing schemes. Fingered Speedwell is a Red Data Book species and currently classified as Endangered. In the UK it is found only in Breckland, where it survives at a handful of locations that are specially managed for rare plants, although it is worth keeping an eye out for elsewhere – occasionally it even turns up as a garden weed in Thetford!

Of the trio of rare Breckland speedwells of disturbed and cultivated ground, this is the most distinctive in appearance. It is an annual plant with stems that may grow to 20 cm in height, but it is usually much smaller. The lower leaves are very distinctive, being deeply cut to form a series of three to seven fingers. The flowers are a rich deep blue and the green seedpods often have a bluish tinge to them. This is often the earliest of the rare speedwells to flower and may sometimes by found in bloom as early as mid-March.

Arable

Small Alison
Alyssum alyssoides

This little plant was once included in the Red Data Book as a rare native species, but these days is generally considered to have been introduced from mainland Europe. Although it is occasionally found in other parts of the UK, it has always been most persistent in East Anglia and thoroughly deserves its place as a Breckland star species. Sadly it seems to have been lost from its former sites in Norfolk, but can still be found on arable margins in the southern parts of the Suffolk Brecks. However, knowledge of its appearance may allow it to be re-found on sandy field edges elsewhere – keep a sharp eye out!

An unobtrusive plant which may grow to 25 cm in height, but is often much shorter. The stems and simple ovate leaves are densely covered in silvery star-shaped hairs, which give the entire plant a greyish look at any distance. The tiny pale yellow flowers appear in tight clusters at the top of the plant, the clusters elongating as the flattened disc-shaped seedpods gradually develop.

Star species

Fine-leaved Fumitory
Fumaria parviflora

The fumitories are delicate plants mostly found as annuals on lightly disturbed ground. This species is fairly widespread on chalky soils in southern Britain, but it is nowhere common and officially classed as "Scarce" in the UK. In Breckland it occurs at scattered locations on disturbed chalky ground where this merges into the open landscapes of the Fens.

When growing among other plants, this species develops a scrambling habit, its thin and slightly fleshy stems scrambling over its neighbours. The whole plant has a slightly blue-green look to it and the narrow finely cut leaflets are clearly grooved on the upper side. Most of the fumitories have rich pink flowers, but this species has very pale pink, almost white, flowers with a dark maroon tip.

Similar species: Compare with the pink-flowered Common Fumitory on page 62.

Common Poppy
Papaver rhoeas

Poppies enjoy great familiarity through their association with Remembrance Day, after poppies carpeted the ground disturbed by the trenches in WW1. The Common Poppy still appears commonly on cultivated field edges where left unchecked by herbicides, although, like all arable 'weeds', it continues to decline. Most poppies you find will be of this species, recognised by its broadly overlapping petals and hairless seedpod, which is a little longer than it is wide.

Rough Poppy
Papaver hybridum

A poppy of chalky soils and most likely to be found out on the western margins of the Brecks, towards Newmarket and west of Weeting and Mundford. It is readily recognised as a poppy, but has more finely cut lobes to its smaller leaves and also smaller flowers, which are typically only 5 cm or less across. The seed capsule is similar in shape to that of Common Poppy, but in this species it has coarse bristles on the sides.

Prickly Poppy
Papaver argemone

The rarest of the poppies in Breckland and most likely to be found on the western edges of the area, where chalk comes close to the surface and has not been overlain by drifting sands. It may grow to 30 cm in height but is often much shorter, especially on dry soils. The flowers are typically around 5 cm in diameter and have obvious gaps between the petals. The seed capsules are much longer than wide and have stout bristles on them.

Long-headed Poppy
Papaver dubium

After Common Poppy, this is the most widespread of the poppies in Breckland and may be found in small numbers throughout the area on cultivated ground, often mixed in with larger numbers of Common Poppy. The flowers are typically smaller than the largest of the Common Poppy flowers, but the two do overlap in size. The petals of this species overlap at the base and the seed capsule is much longer than wide, with a smooth surface.

Yellow-juiced Poppy
Papaver lecoqii

For a long time, this species was considered to be just a variant of the Long-headed Poppy, which it closely resembles, but it is now accepted as a separate species. It is rather rare in Breckland and most likely to be found on the western edges of the area, becoming more common in the Fens. Its petals typically have small gaps between them and the seed capsule resembles that of Long-headed Poppy, but it differs in the yellow (not white) sap – break off just a small section of leaf to check.

Common Fumitory
Fumaria officinalis

A widespread species and common on the chalkier soils of the cultivated parts of the Brecks. Plants start as low mounds of pale blue-green, finely cut foliage. These early growths produce stems that may reach 50 cm or so in length and which bear spikes of flowers in the leaf axils. Fumitory flowers have a rather odd tubular structure, with the pale pink petals tipped in a darker maroon colour. The flowers are followed by small, rounded seedpods.

Similar species: Compare with Fine-leaved Fumitory on page 60.

Arable

Field Pansy
Viola arvensis

Pansies are very familiar and popular as garden plants, with cheery little 'faces' on their petals. This species is much smaller than the big, blousy plants of gardens, with flowers that are typically less than 2 cm across. These are usually whitish or cream with a yellow centre, but may occasionally also have traces of violet on them. A common species in arable fields and other cultivated areas.

White Melilot
Melilotus albus

A fairly common plant found on all kinds of cultivated land, especially on rough corners, waste areas and roadsides. Plants bear leaves with three leaflets, rather resembling those of clovers but larger; upper leaves are narrower than lower leaves. Stems may grow to nearly two metres in height and are many-branched. The narrow flowers are similar to those of the clovers, but are carried in slender upright spikes.

Ribbed Melilot
Melilotus officinalis

Another common plant of cultivated and disturbed ground and often found along field edges and in weedy corners. The three-fingered leaves are a little narrower than those of White Melilot and usually more sharply toothed on the margins. The narrow, bright yellow flowers are carried in long narrow spikes at the top of branching stems, which may grow to 150 cm in height. Flowers are followed by rows of oval seedpods with ribbed surfaces.

Sun Spurge
Euphorbia helioscopia

A common species on chalkier soils and found throughout much of Breckland on cultivated ground, including gardens. An unassuming plant with little about it to catch the eye. Stems may grow to 30 cm in height and are single or only with a few branches. The leaves are broadly rounded with blunt tips and regularly toothed along the margins. The tiny flowers (you might need a hand lens!) are yellowish-green and appear in tight clusters at the tops of the leafy stems.

Petty Spurge
Euphorbia peplus

A very common rather 'weedy' plant of all kinds of cultivated and disturbed ground, often forming spreading colonies of plants on gravel drives. Stems grow to around 30 cm in height and are often multi-branched, with the lower branches bare and resembling miniature trees. The tiny flowers (use a hand lens!) nestle among leafy bracts. Leaves and bracts are narrower than those of Sun Spurge.

Scarlet Pimpernel
Anagallis arvensis

A common species on most kinds of disturbed ground on farmland and in urban areas. The plants are smooth and hairless and have many-branched, square-sided stems that spread low to the ground. Leaves are oval and in opposite pairs on the stems. The bright orange-red flowers open in sunny weather but usually remain closed when it is cloudy.

Arable

Field Penny-cress
Thlaspi arvense

A widespread species of cultivated ground which may be found as odd single plants, but can also occur in large quantities on richer, heavier soils. The leaves and stem are hairless and bright green, the leaves being wavy-edged, unlobed and clasping the stem at their bases. The small, white, four-petalled flowers appear at the top of the stems (which grow to 50 cm in height) and are followed by broadly rounded, disc-shaped seed capsules.

Field Pepperwort
Lepidium campestre

This is a rare plant in the Brecks, but is similar to Field Penny-cress and so care should be taken to look for the differences that separate the two species. Field Pepperwort differs from Field Penny-cress in being densely covered in short white hairs. The leaves clasp the stem and are narrower and are more sharply toothed on the edges. The seed capsules are narrowly spoon-shaped.

Similar species: Smith's Pepperwort (*L. heterophyllum*) has seed capsules with a prominent spike at their tip.

Hoary Cress
Lepidium draba

This species was originally introduced from the Mediterranean region but is now widespread in the UK, although it is rather scarce on the thinner soils of central Breckland. The simple leaves with toothed edges and clasping bases are similar to those of the pepperworts, but this is a perennial species that spreads below ground to form creeping colonies of stems. The small four-petalled flowers appear in rounded 'mop heads' and are followed by rounded, unwinged seed capsules.

Common Swine-cress
Lepidium coronopus

Far from being the showiest plant in the area, Common Swine-cress grows on richer arable soils and is widespread on disturbed ground. Plants are hairless, dull green and form flat rosettes of irregularly lobed leaves. The tiny flowers are four-petalled, but often rather irregularly shaped and tightly clustered together in the centre of the plant. They are followed by rough and knobbly seed capsules.

Lesser Swine-cress
Lepidium didymum

An introduced species that has become well established on disturbed ground and as a weed of pavements and gardens. It rather closely resembles Common Swine-cress, but spreads to become a many-branched plant that creeps over the ground to form mats of finely cut leaves. Stems have white hairs and the petals are less than 2 mm in length, or even absent, leaving greenish-looking flowers in dense clusters.

Shepherd's-purse
Capsella bursa-pastoris

A very common annual plant, which may be found in almost any kind of disturbed habitat and especially along roadsides and in urban areas. The first leaves appear as a flat rosette and are very deeply cut into pointed lobes. The upright stems grow from the leaf rosettes, up to 40 cm in height, but often much shorter. Stem leaves are unlobed. The flowering stems elongate as the flowers mature and carry very distinctive heart-shaped seed capsules.

Arable

Hairy Bitter-cress
Cardamine hirsuta

An abundant small species found in almost all types of habitats and often one of the first species to appear on bare, disturbed ground. The leaves are pinnately lobed, with plants starting as a basal rosette of leaves. Small white flowers are clustered at the stem tops, with each flower having six stamens in the centre (two shorter than the others). Seed capsules are long, thin and carried upright.

Similar species: Wavy Bitter-cress (page 21) is very similar, but the flowers have six stamens.

Thale Cress
Arabidopsis thaliana

This plant competes with Hairy Bitter-cress as one of our commonest plants, but is a little more restricted to arable and urban areas and less likely in shady spots. At first glance, it rather resembles the bitter-cress species but the leaves of the basal rosette are not pinnately lobed – they are simple, wavy-edged and roughly hairy. Plants have a smooth blue-green look to them and the long and thin seed capsules are held out at an angle, well away from the main stem.

Whitlow-grass
Erophila verna

Often found in colonies of many plants packed closely together, this is one of Breckland's smaller species (usually to 10 cm or less in height) and so can be missed easily. It can be found on a variety of cultivated and disturbed areas, as well as on mossy walls and pavements. The tiny leaves are all in a basal rosette and are oval with toothed edges. The flowers are carried in clusters, well above the leaves, and have four very deeply notched petals.

Wild Radish
Raphanus raphanistrum

A common species of cultivated ground and field margins. It is usually rather shorter than the other brassica family species of field margins, but typically well-branched and forming low, bushy plants to 60 cm in height. The coarse and bristly leaves are deeply pinnately lobed. The flowers may be white or pale yellow, with widely separated petals, typically with darker veins. Seed capsules are chunky, with constrictions between each of the seeds inside.

Annual Wall-rocket
Diplotaxis muralis

A plant of dry, stony or disturbed places and a common species in urban areas as a weed of roadside cracks and garden walls. This is one of the smaller yellow brassica family members, usually growing no more than 30 cm in height. Plants have a basal rosette of smooth and rather irregularly lobed or toothed leaves. Flowers are carried in a short cluster at the top of the stems, with slender seed capsules carried on rather short stalks.

Perennial Wall-rocket
Diplotaxis tenuifolia

Mostly a coastal plant in East Anglia, this is a rare species in Breckland but increasing as an urban plant and in disturbed soil around Thetford and in the Suffolk Brecks. A much larger plant than Annual Wall-rocket, with a bushy, much-branched habit and growing to 80 cm in height. Leaves are pinnately lobed and the slender seed capsules have rather longer stalks than those of the annual species. The flowers are sweetly scented.

Arable

Charlock
Sinapis arvensis

Members of the cabbage family, with their four-petalled yellow flowers, are common and numerous and often found along arable margins. The chalky disturbed soils of the arable parts of Breckland are especially favoured by this species. It has lower leaves that typically have a large end lobe and a short series of small side lobes; upper leaves are simple and clasp the branched stems. Flowers have sepals that are angled away from the stems and seed capsules have stout 'beaks' on their tip.

White Mustard
Sinapis alba

Generally rather rare in Breckland, but populations of this species come and go and it can be frequent in arable areas for a few years if seed is spilt after a crop has been grown for mustard seed. It resembles Charlock but has pinnately lobed leaves and the stem leaves are stalked, not clasping the stem. The sepals are held angled away from the stems and the seed capsules are hairy with very long, sabre-like 'beaks' at the end.

Oil-seed Rape
Brassica napus subsp. *oleifera*

Now very commonly grown as a crop to provide seed for making various vegetable oil products, spilt seed from commercial and agricultural vehicles has made this by far the commonest yellow brassica species, especially along the sides of major roads and arable margins. The lower leaves have long stalks, while the upper leaves are stalkless; the entire plant has a blue-green bloom to it. Flowers and seedpods are larger than those of most other brassica species in the area.

Flixweed
Descurainia sophia

A species that seems to have been introduced into the UK by early settlers from southern Europe. Common on light sandy soils across most arable parts of the Brecks, this is a very delicate plant that may not be immediately recognised as a member of the brassica family. Stems are stiffly upright and may grow to 80 cm or so in height, bearing very finely cut leaves. The pale yellow flowers are clustered at the top and have tiny petals.

Tall Rocket
Sisymbrium altissimum

Although originally introduced from mainland Europe, this species has become something of a Breckland speciality and can be found commonly on rough corners of arable fields and other disturbed areas. Stems are well branched lower down and may grow to a metre in height. The leaves are very deeply cut, almost to the veins and giving a ladder-like effect. The flowers are small and pale yellow in colour, and are followed by long, slender seed capsules. Old dry plants can blow around like tumbleweed, providing effective seed dispersal.

Hairy Rocket
Erucastrum gallicum

A rare introduced plant, found on disturbed ground mostly around Brandon, Mildenhall and Lakenheath. The lower leaves are rather broad and are deeply lobed, each side lobe being further more finely lobed along its margin. Upper leaves become progressively smaller up the stem and are deeply lobed. The flowers are pale lemon yellow, while the whole plant is covered in mealy hairs, which tell it apart from similar species.

Redshank
Persicaria maculosa

An abundant plant of all kinds of cultivated and disturbed places and especially common in damper soils, this is an easily recognised species once known, even if a little variable in appearance. Plants may grow from a few centimetres to almost a metre in height (largest on old muck heaps!) and the reddish stems have simple pointed leaves, each with a dark mark on them. Tiny flowers are crowded in a thin 'rat tail' spike and may be white or deep pink. Flower spikes have no glandular hairs at the base.

Pale Persicaria
Persicaria lapathifolia

Quite common in wet or disturbed ground in the river valleys and along the western edge of the Brecks, becoming more common in the Fens. Very similar to Redshank in appearance and sometimes showing the dark leaf blotch. The white or pale pink flowers are crowded in long, often curved, spikes. Most easily told from Redshank by the flower spikes, which have yellowish glandular hairs at the base of the spike and often on the backs of the flowers also.

Common Knotgrass
Polygonum aviculare

An abundant plant found in bare and disturbed places, and often appearing in great quantity in stubble fields after harvest. Wiry stems trail across the ground and plants overlap to form complex mats with simple, lance-shaped leaves, those on the main stems being much larger than those on side stems. Tiny white flowers appear in the leaf axils and are easily missed.

Similar species: Equal-leaved Knotgrass (*P. arenastrum*) is common on well-trodden ground. It has ovate leaves that are all more or less the same size.

Black Bindweed
Fallopia convolvulus

A common species of cultivated ground and a strange little plant, as it seems to want to twine like a climber, but usually ends up sprawling across the ground! Wiry, usually reddish, stems scramble over nearby vegetation and bear arrowhead-shaped leaves with slender tips. The tiny white flowers grow on a slender stem from the upper leaf axils and are followed by pale green, hanging seed capsules.

Field Bindweed
Convolvulus arvensis

A strong-growing, perennial species that can form extensive patches of trailing stems in all kinds of disturbed ground. Twining stems have arrowhead-shaped leaves up to 5 cm in length, with blunter tips than those of Black Bindweed. The bell-shaped flowers are borne in great abundance along the stems and may be white or striped with pink.

Similar species: Compare with the larger bindweeds of hedgerows (see Part One).

White Bryony
Bryonia dioica

This species tends to like hedges in open farming areas and can also sometimes be found on cultivated ground. The leaves are deeply palmately lobed and grow on stems that climb strongly into and over surrounding vegetation, using spiral tendrils that grow from the leaf axils. Male and female flowers appear in clusters on separate plants and, in female plants, are followed by showy clusters of dull red berries.

Arable

Sticky Mouse-ear
Cerastium glomeratum

A common species of cultivated ground, this plant gets its name from its broadly rounded, furry leaves. The stems grow up from a basal rosette of pale green leaves, with all parts of the plant covered in glandular hairs that are slightly sticky in character. The five-petalled flowers are carried in tight clusters at the top of the stems and the leafy bracts below them are all-green, without papery edges.

Similar species: Compare very carefully with other mouse-ears (see Part One).

Common Chickweed
Stellaria media

A very common plant found growing in almost all types of habitat, although particularly common on disturbed and cultivated soils. Very variable, it closely resembles the mouse-ears, but is more or less hairless, the stems having a distinctive single line of hairs down one side. Leaves are ovate or narrowly heart-shaped on narrow leaf stalks.

Similar species: Compare with Lesser Chickweed (see Part One).

Corn Spurrey
Spergula arvensis

Although declining nationally, this species can still be found quite commonly on the sandy soils of Breckland, especially on field margins and corners where herbicides have not reached. Young plants first appear as clusters of grass-like, but succulent, linear leaves. Stems elongate and become sprawling, with the white flowers appearing in many-branched, open heads.

Similar species: Other spurreys have similar leaves but pink flowers (see Part One).

Night-flowering Catchfly
Silene noctiflora

A scarce species that may be found on the lighter chalky and sandy soils of the Brecks, most often among broad-leaved crops such as Sugar Beet. It greatly resembles White Campion, but tends to have slightly narrower leaves and flowers that roll up their petals during the day, opening them again in late afternoon or evening. The petals are very pale pink on top, with the backs (seen when rolled up) having an odd olive-yellow cast to them.

White Campion
Silene latifolia

A very common and easily recognised plant that can be found in all kinds of disturbed and waste places along field margins, roadsides and in urban areas. The simple, narrowly oval leaves and upright branched stems are covered in downy hairs. The pure white flowers are carried in branched heads at the top of the stems, the flowers themselves having a strongly ribbed, swollen basal section.

Cut-leaved Crane's-bill
Geranium dissectum

A widespread and common species of field edges, hedge bottoms and roadsides, especially where the ground has been disturbed. A low-growing, many-branched plant that has leaves deeply cut almost to the veins. The flowers appear rather small for the size of the plant, opening less widely than those of other crane's-bills.

Similar species: Compare with other crane's-bills in **Part One**.

Arable

Broad-leaved Willowherb
Epilobium montanum

The willowherbs are a difficult group of plants that are all rather similar and can require technical scrutiny to identify. This is one of the more common species of disturbed, cultivated and urban places. Its leaves are broadly rounded at the base, stalkless and clasp the unridged stem. The small flowers have a stigma in the centre that is very clearly divided into four lobes, forming a cross at the top.

American Willowherb
Epilobium ciliatum

A relative newcomer to our country, but now very widespread on cultivated ground on field margins and in urban areas. The stems have four ridges along their length and bear slightly glossy, narrow leaves with toothed margins and a short leaf stalk. The tops of the plants have both glandular and non-glandular hairs on them. The stigma at the centre of the flowers is club-shaped and not divided into a cross.

Short-fruited Willowherb
Epilobium obscurum

An uncommon species, found more often in damp or shady disturbed places. This species is best described as being intermediate between the two species above. It has four-ridged stems, narrow leaves, glandular hairs and club-shaped stigmas like American Willowherb, but the leaves are more or less without stalks, like those of Broad-leaved Willowherb.

Similar species: Square-stalked Willowherb (*E. tetragonum*) is very similar, but the top of the plant is covered with flattened, white hairs.

Small Nettle
Urtica urens

There can be few people who do not recognise a nettle when they see one, after a nasty shock as a child! Small Nettle is a low-growing, annual species of cultivated and disturbed ground, with the same 'bite' as the perennial Common Nettle. Compared with the latter, it has more oval-shaped leaves with shorter leaf stalks and its flowers are clustered tightly among the leaf bases, not hanging like small catkins.

Annual Mercury
Mercurialis annua

An uncommon species in Breckland, which is most likely to be found on disturbed ground in and around Thetford, especially in gardens. The simple leaves are smooth and glossy with lightly toothed margins. Plants are well-branched and grow to around 50 cm in height. The flowers are small and green and therefore easily overlooked. Male and female flowers appear on separate plants, the male flowers being more conspicuously carried on upright spikes.

Fat-hen
Chenopodium album

This plant is abundant in cultivated soils, but may not catch the eye as its flowers are peculiar tiny green structures carried in dense clusters towards the top of the plant and covered in a white mealy coating. This variable species ranges from 20 cm to over two metres in height if growing in enriched soil! The leaves are broadly shaped like spearheads, with irregularly toothed margins.

Arable

Spear-leaved Orache
Atriplex prostrata

Originally a plant of saltmarsh margins, this species has spread inland along salt-treated road edges and is now found frequently as a weed of muck heaps and waste ground on field borders. The leaves are spear-shaped and usually have two obvious lobes at their bases. The tiny green flowers are clustered tightly in short spikes in the leaf axils. Difficult to tell apart from Common Orache, but the seed capsules have triangular outer segments that are fused together at the bottom for only a quarter of their length.

Mike Crewe; Paul Sterry / Nature Photographers

Common Orache
Atriplex patula

A common species of roadsides, disturbed ground and arable margins. This is a rather straggly, sprawling plant that has narrowly spear-shaped lower leaves with two pointed lobes at the base, while upper leaves tend to be narrower and unlobed. Difficult to tell apart from Spear-leaved Orache, but the seed capsules have broadly diamond-shaped outer segments that are fused together at the bottom for half their length.

Mike Crewe

Grass-leaved Orache
Atriplex littoralis

Originally a plant of saltmarshes, but has spread inland thanks to the salting of roads during the winter and can now be found along roadsides and on bare or cultivated ground. The flowers are small and green, clustered tightly together in short spikes in the leaf axils. The leaves are longer and narrower than those of other oraches, with coarsely toothed margins.

Mike Crewe

Fig-leaved Goosefoot
Chenopodium ficifolium

Goosefoots are common annual plants of cultivated and disturbed ground and may be found throughout Breckland. This species has rather narrow leaves that are coarsely toothed on the margins and have two larger lobes at their bases. The plants tend to be rather narrow and stiffly upright, with a few lower branches and growing to around 90 cm in height. The tiny green flowers are clustered in tight spikes near the top of the plant.

Red Goosefoot
Oxybasis rubra

This species can be found in a range of disturbed habitats, but seems to favour the enriched soils of muck heaps and pig farms, where stems can grow to nearly two metres in height! It often has a reddish tinge to its tightly packed spikes of tiny flowers. The pale-veined leaves have a rather thick, fleshy look to them and have shiny surfaces and deeply toothed or lobed margins.

Maple-leaved Goosefoot
Chenopodiastrum hybridum

An uncommon and much-branched species of field edges and areas of disturbed ground on lighter soils. The tightly packed clusters of flowers appear on rather open, branched stems towards the top of the plant. The lower leaves are broadest at the base, with a few large pointed lobes on their margins, giving the appearance of a maple leaf. Upper leaves are narrower with relatively long leaf stalks.

Arable

Indehiscent Amaranth
Amaranthus bouchonii

The amaranths are a tricky group of plants that largely originate from Central and South America. Long-stalked, oval leaves have deeply marked veins and plants grow in great abundance in broad-leaved crops such as Sugar Beet. Flowers are tiny and green, densely packed into narrow spikes. Telling the various species apart mostly involves detailed study of the minute flower parts under magnification!

Similar species: Common Amaranth (*A. retroflexus*) and Green Amaranth (*A. hybridus*) are scarce, but similar and require close scrutiny.

Common Cleavers
Galium aparine

Also known as Goosegrass, Sticky Buds or Sticky Willy, this plant is much loved by schoolchildren for its ability to cling to clothing and annoy friends! Its four-angled stems and whorled leaves are armed with hooked hairs and it uses these to scramble over nearby vegetation. The greenish-white flowers are small and tucked amongst the leaves, being followed later by round fruits with hooked hairs.

Similar species: Compare with other bedstraws on page 26 and in **Part One**.

Field Madder
Sherardia arvensis

This low-growing bedstraw relative can be found on dry, usually sandy soils on field margins and tracksides in arable areas. The stems spread out to form low mats of vegetation and the finely pointed leaves are borne in whorls of four to six on the stems. The small flowers appear in bunched clusters at the ends of the stems and are pale lilac in colour.

Field Bugloss
Anchusa arvensis

A widespread and common species of arable margins and cultivated ground, this is a well-branched plant growing to around 50 cm in height. The leaves are simple and lance-shaped with strongly undulating margins and the whole plant is covered in stiff, white, bristly hairs. The bright blue flowers with tubular bases and white centres grow in crowded clusters among bristly bracts.

Common Fiddleneck
Amsinckia micrantha

Known as Tarweed in its native North America, this plant seems to have been accidentally introduced to the area and is now common on dry sandy soils in cultivated areas throughout much of Breckland. The leaves are simple and narrow, without leaf stalks and with wavy edges, and the whole plant is covered in stiff bristly hairs. The little yellow flowers arise at the ends of the stems in curled spikes and gradually open just one or two at a time.

Field Forget-me-not
Myosotis arvensis

A common plant of disturbed or cultivated ground. The stems arise from a basal rosette of simple bristly and hairy leaves and grow to a height of 30 cm, or often much less. The tiny bright blue flowers open on an unfurling spike, one or two at a time. Small seed capsules develop stalks that are longer than the capsules.

Similar species: Early Forget-me-not (see **Part One**) is smaller and has seed capsules on stalks shorter than the capsules.

Arable

Black Nightshade
Solanum nigrum

A widespread and quite common species of richer soils in areas of cultivation, especially gardens, field edges and old muck heaps. The plants are many-branched and have simple, broadly ovate leaves with entire or lightly toothed margins. The star-shaped flowers grow in small branched clusters in the leaf axils and are followed later by green berries that eventually turn shiny black.

Green Nightshade
Solanum physalifolium

An introduced species that is generally scarce in East Anglia but can be rather common on arable margins in Breckland. Plants grow to around 40 cm in height and are typically well branched. The simple, broadly ovate leaves have strongly toothed margins and the whole plant is covered in fine hairs. White flowers are followed by green berries, which develop a tinge of olive-brown but do not turn black.

Small Nightshade
Solanum triflorum

This little plant was introduced from North America and can be found on rough field margins and rough corners in a few areas of the Suffolk Brecks, particularly around Icklingham and Mildenhall. It has the same white and yellow star-like flowers typical of the nightshades, in clusters of three, and followed by green berries with a marbled pattern to them. It differs from other nightshades in the very deeply cut and lobed leaves.

Common Field Speedwell
Veronica persica

Although a relatively recent introduction to the UK from the European mainland, this species has become a very common and widespread plant in all kinds of disturbed and cultivated places. The plants can be quite well branched and spreading, with ovate, well-toothed leaves that clasp the stem at their base. Flowers are very bright blue, the lower petal being smaller and usually (although not always) paler than the other petals. Seed capsules are two-lobed, the slightly pointed lobe tips diverging by 90 degrees or more.

Green Field Speedwell
Veronica agrestis

A rather uncommon species of field margins and cultivated places, this plant was once more common but has declined greatly in recent years. In growth it is very similar to Common Field Speedwell, although generally smaller in all of its parts. The small flowers (up to 8 mm across) are typically pale blue, sometimes pale lilac or whitish. The two-lobed seed capsules are rounded and carried between rather long and narrow sepals.

Grey Field Speedwell
Veronica polita

As with the Green Field Speedwell, this plant was once more common but has declined in recent years and is now a scarce species of field margins and cultivated places. It is very similar to Common Field Speedwell, although generally smaller and the leaves have a dull tone. The small flowers (up to 8 mm across) have all petals the same shade of deep blue. The two-lobed seed capsules are rounded and carried between relatively short and broad sepals.

Arable

Wall Speedwell
Veronica arvensis

This tiny plant occurs in a wide range of habitats. It is common on field margins and as a weed in gardens, but may also be found on bare patches in grassy areas and growing on walls and cracked pavement. The lower part of the plant has small ovate leaves with toothed margins, while the flowers are carried in dense spikes from the upper leaf axils. The tiny flowers (just 2–3 mm across) have very short stalks and are tucked well in amongst the long-pointed leafy bracts.

Ivy-leaved Speedwell
Veronica hederifolia

A widespread and common species that comes in two forms: one grows in shady places in woods and hedge bottoms, the other is commonly found on all kinds of cultivated and disturbed ground. Plants form elongated creeping stems that bear three- to five-lobed leaves that rather resemble those of Ivy. The flowers are pale blue or lilac and appear singly on long stalks in the leaf axils.

Corn Mint
Mentha arvensis

A rather scarce plant in Breckland, favouring damp soil on field edges but also sometimes found in bare places in woodland. A low-growing plant, although the flowering stems may reach 60 cm in height. Stems and the oval-shaped opposite leaves are all covered in dense hairs and the whole plant has the distinctive smell of mint. Pale pink flowers are tightly clustered at the leaf bases and stems are topped with small leaves.

Similar species: Water Mint (page 30) has stems topped with a cluster of flowers.

Henbit Dead-nettle
Lamium amplexicaule

A common species of dry soils in a variety of cultivated and disturbed places, as well as on the gravelly edges of grassy heaths. The paired hairy leaves are stalkless and their broad, rounded bases clasp the stem and encircle it like a collar. The pale purple flowers have long tubular bases and stand almost upright at the top of the plant.

Red Dead-nettle
Lamium purpureum

A very common plant that grows on all kinds of cultivated and disturbed ground and is common in towns and gardens. It germinates in the autumn and often flowers late in the year, appearing again in late winter or early spring. The short clumps of hairy leaves and stems are often purple-tinged towards the top, and the narrow, tubular-based flowers appear in small clusters, tucked among the leaves.

Cut-leaved Dead-nettle
Lamium hybridum

This species mostly prefers richer, heavier soils and is most likely to be found on field edges around the margins of Breckland, being largely absent from the sandy conditions of the central area. It is very similar to the very common Red Dead-nettle, but differs in having leaves and flower bracts that are more deeply lobed or toothed around their margins.

Arable

Common Hemp-nettle
Galeopsis tetrahit

A fairly common species of disturbed ground, especially in damper areas along river valleys. Plants may grow to around 90 cm in height, although often much less, and the square-sided stems bear narrowly nettle-like leaves in pairs – but without the sting! The flowers appear towards the top of the plant, arranged in clusters in the leaf axils. The flowers are variable and may be pink or white, but usually have a prominent yellow patch with purple veins on the lower lip.

Bifid Hemp-nettle
Galeopsis bifida

Less frequent than Common Hemp-nettle, but careful checking of the flowers may often reveal the two species growing together. Leaves are narrowly nettle-like (without stings). The flowers appear towards the top of the plant, arranged in clusters in the leaf axils. They are typically a little narrower than those of Common Hemp-nettle and the lower lip has less yellow, more purple, and is notched at the tip.

Venus's-looking-glass
Legousia hybrida

This cheerful little plant occurs along field margins on chalkier soils and is still quite regularly found in suitable places in Breckland. Stems may grow to around 30 cm in height, but are often much shorter. The simple leaves are wavy-edged and stalkless. The rich purple petals are small – around 1 cm across – and surrounded by the longer green sepals. Good weather is needed to see this species at its best, as the flowers only open on sunny days.

Small Toadflax
Chaenorhinum minus

A tiny plant that may be easily missed without careful searching. It may be found on light sandy or chalky cultivated soils as well as on gravelly areas such as roadside banks, car parks and dirt tracks. Plants are typically less than 25 cm in height and well branched. The stems bear simple narrow leaves and the whole of the upper part of the plant is covered in fine glandular hairs. The pink and white flowers are like little snapdragons.

Tansy-leaved Phacelia
Phacelia tanacetifolia

Also known as Bee Balm, this species is a relative newcomer to Breckland and occasionally grown as a temporary cover crop or in strips as a pollinator plant for bees; odd plants may turn up on field margins from spilt seed. The purple-blue flowers with long stamens grow in curled heads, which slowly unfurl as the flowers open. The leaves are deeply and finely cut into a filigree of leaflets.

Cornflower
Centaurea cyanus

The brilliant blue flowers of this species were once a common sight in arable crops, but it is now more likely to be found where it has been sown as part of a 'wildflower' mix. The stiff upright stems grow to 90 cm in height. The lower leaves are deeply lobed, the upper leaves simple with all leaves and stems covered in dense, flattened hairs. Flowers are borne in tight heads at the stem tops and are typically blue, but may occasionally be pink or white.

Arable

Gallant Soldier
Galinsoga parviflora

An introduced plant from South America that is increasing in the area as a garden weed and marginal plant of disturbed and cultivated soil. Plants have branching stems and typically grow to around 60 cm in height. The flowerheads are daisy-like with a yellow centre and white outer segments, but they have just five white petals rather than many. Plants have a few scattered hairs on the leaves and stems. Very difficult to tell from Shaggy Soldier without flower dissection.

Mike Crewe

Shaggy Soldier
Galinsoga quadriradiata

As with Gallant Soldier, this species is slowly increasing in the Brecks as a weedy plant of cultivated and disturbed ground, especially in urban areas. It is very difficult to tell from Gallant Soldier and accurate identification requires magnified inspection of the flowerheads (consult a detailed flora). However, this species tends to be more hairy overall and the flowerheads have broader white petals with narrower gaps between them.

Mike Crewe

Groundsel
Senecio vulgaris

One of the commonest plants in the UK on disturbed and cultivated ground and surely familiar to anyone who has ever done any gardening! Related to the ragworts, it has similarly deeply cut and wavy-edged leaves. The tight flowerheads have no outer petals and consist of a bundle of yellow petalless florets. Flowers may be found in any month of the year.

Mike Crewe

Creeping Thistle
Cirsium arvense

A common plant in open places on both enriched grassland and cultivated ground, this plant is the bane of many a farmer! It spreads by means of persistent underground rhizomes to form extensive patches of narrow prickly leaves and prickly stems. The flowerheads are distinctly lilac-coloured and paler than those of other thistles.

Similar species: Compare with other thistles on page 32 and in **Part One**.

Canadian Fleabane
Conyza canadensis

Also known as Horseweed, this plant arrived in the UK from North America and has spread to become a very common species of field edges and disturbed ground. Plants begin with a basal rosette of pale yellow-green leaves that are hairy and have irregularly toothed margins. The stiff, upright stems can grow to a height of over two metres, but they are often much shorter. The flowerheads are tiny, but appear in great number on an elongated spike. The petals barely reach beyond the green outer bracts.

Guernsey Fleabane
Conyza sumatrensis

An introduction from South America (despite the name!), this species arrived in the Brecks much later than Canadian Fleabane but has quickly become established as a weed, especially favouring urban habitats. The lower leaves are broader and more coarsely toothed than those of Canadian Fleabane and the whole plant has a grey-green look to it. The tightly clustered flowerheads have even shorter petals, and the clusters are broader at the base and more flask-shaped.

Arable

Broad-leaved Cudweed
Filago pyramidata

Always rare in Breckland, this species is known currently from a single location in Suffolk and is unlikely to be found by the casual observer or visitor; however, it closely resembles other cudweeds and it would be an exciting find if rediscovered elsewhere! It resembles Common and Red-tipped Cudweeds, but has yellowish flower bracts and leaves that are broadest towards the tips.

Pineappleweed
Matricaria discoidea

A very common plant of all kinds of open and disturbed places, most often found in gateways, along tracks and other well-trodden places, and also a regular sight in urban areas. The leaves are very finely cut, almost to the veins and the plants are much branched. The yellow flowerheads have no outer petals to them and the whole plant has a surprising smell of pineapple if crushed.

Smooth Cat's-ear
Hypochaeris glabra

While most of the 'dandelion-type' flowers are found in grassy or heathy places, this species typically occurs on disturbed and often gravelly or sandy ground. It is a small plant, growing to no more than 40 cm, with a basal rosette of deeply toothed and completely hairless leaves. The flowerheads are no more than 1.5 cm across and often do not open fully on cloudy days with little sun.

Similar species: Compare with cat's-ears and hawkbits in Part One.

Smooth Sow-thistle
Sonchus oleraceus

A very common and widespread plant found in all kinds of cultivated ground. Sow-thistles are like dandelions on stilts, having dandelion-like flowers in open heads at the top of tall stems. This species has blue-green leaves with a slight bloom and lobed margins. The leaf bases clasp the stem and terminate in a long point on either side. The flowerheads may be either all yellow or paler and cream-coloured or bluish around the edge.

Prickly Sow-thistle
Sonchus asper

Another very common sow-thistle found in all kinds of cultivated, disturbed and urban places. Stems grow to around a metre in height and bear leaves with a glossy shine and prickly edges, rather like thistle leaves. The leaf bases clasp the stem and terminate in a rounded lobe on either side, each with a row of spikes, all of more or less equal size. The flowers are carried in clusters in open branched heads.

Perennial Sow-thistle
Sonchus arvensis

A common species of open places in all but the drier soils of the heathlands. As a perennial species, it spreads by underground rhizomes and can be a persistent plant in arable land, especially in Sugar Beet fields. The stems may grow to over a metre in height and the leaves are long and lance-shaped with coarsely toothed margins and a long-pointed tip. The flowerheads are relatively large – up to 5 cm across – and have bright yellow glandular hairs on their outer bracts.

Arable

Scentless Mayweed
Tripleurospermum inodorum

Few areas of disturbed or cultivated ground seem to be without one of the mayweeds. This species is by far the commonest in Breckland and has leaves with little or no scent when crushed. They are very finely cut, almost to the veins, and the stems are many-branched, forming bright green leafy mounds of vegetation. Apart from the lack of scent, this species can be told from Scented Mayweed by its slightly domed flowerheads, which are solid at the base.

Scented Mayweed
Matricaria chamomilla

This species is less common than Scentless Mayweed in the Brecks, avoiding the drier soils and being rather more local in the damper river corridors. Unlike its close relative, it has a rather pleasant aroma to the leaves when crushed. The stems and leaves of the two mayweeds are very similar and difficult to tell apart, but Scented Mayweed has flowerheads that are more pointed in profile and which have hollows at the base inside.

Corn Chamomile
Anthemis arvensis

Once quite common as an ancient introduction from mainland Europe, this species has decreased greatly in recent times but may still be found on arable margins in Breckland and appears regularly in sown 'wildflower' mixes. The daisy-like flowers are very like those of the mayweeds, but the whole plant has a pleasantly sweet smell and the stems and leaves are covered in flattened whitish hairs.

Fool's Parsley
Aethusa cynapium

Plants of the umbellifer family can be rather daunting and 'samey' in appearance, but only a few are likely to be found on cultivated ground. This species is quite widespread in arable and garden soils and can grow to a height of almost a metre, although it is often much shorter. The stems and leaves are completely hairless and the flowerheads have very distinctive bracts that hang down beneath (see image). Fruits are round and hairless.

Similar species: Compare with Wild Carrot in Part One.

Ground-elder
Aegopodium podagraria

A very common plant which often invades cultivated ground from nearby hedge bottoms or woodland edges by means of underground rhizomes, spreading to form extensive patches. The leaves appear in April and soon carpet the ground, with their pinnate leaves having rather broad leaflets. The flowers arise in flat heads above leafy stems and are followed by oval, ridged fruits.

Similar species: Compare with other umbellifers on pages 35–37 and in Part One.

Common Cornsalad
Valerianella locusta

This unassuming little plant is rather rare in the Brecks, but may be found around Thetford and especially on dry sandy ground to the northeast of the town. It is hairless and well-branched, with simple wavy-edged leaves that are without leaf stalks and arranged in pairs on the stems. The flowers might look white at first glance, but are in fact a very pale blue when seen close to. Plants produce lots of seed and consequently often grow in low colonies of many individuals.

Walls

Yellow Corydalis
Pseudofumaria lutea

Originally introduced to the UK as a garden ornamental, this species has become well established and may be found growing on sunny walls in many of Breckland's towns and villages. It produces rounded clusters of fleshy stems with finely cut, fern-like foliage. The flowers are bright yellow, tubular in shape and appear in spiked clusters above the leaves.

Wallflower
Erysimum cheiri

Very popular at one time as a bedding plant, this species – as its name suggests – is much more at home when growing from a chalky wall. It may be found in several of Breckland's towns and villages, especially around Thetford. Plants are perennial and form woody stems at the base. Fresh leaves emerge in spring and new growth produces four-petalled flowers of a rich golden yellow.

Ivy-leaved Toadflax
Cymbalaria muralis

This southern European plant has made good its escape from local gardens and can now be found quite commonly on flint and stone walls around the Brecks. The slender stems trail across walls and root at the leaf nodes, wherever they find a suitable spot. The five-lobed leaves are slightly fleshy and resemble small Ivy leaves. Flowers appear throughout the summer and have a two-part outer lip, like little snapdragons.

Walls

Pellitory-of-the-wall
Parietaria judaica

A scarce but slowly increasing species in Breckland that may be found growing from old walls, pavement cracks and other similar places around human habitation. This is a rather something-and-nothing plant that does little to draw attention to itself! The simple leaves are glossy and have deeply impressed veins. Later in the year, reddish flowering shoots grow from the leaf rosettes and carry tiny white flowers that require a hand lens to see fully.

Wall Bedstraw
Galium parisiense

A Nationally Scarce species classified as 'Vulnerable', in East Anglia this is something of a Breckland speciality. It is generally found on chalky soil, but the best-known sites locally are on the walls of ancient buildings, such as the Nunnery at Thetford and at Weeting Castle. Leaves have forward-pointing prickles on their margins and appear in whorls of 5–7, while the tiny greenish flowers often have a slightly reddish tint.

Similar species: Compare with Common Cleavers on page 79

Red Valerian
Centranthus ruber

Although plentiful around East Anglia's coastal towns and villages, this species is less common in Breckland but may still be found growing on walls or sunny banks in urban areas. Stout, woody-based stems produce slightly fleshy, simple leaves in opposite pairs. During summer, large showy heads of tiny flowers are hard to miss and the flowers may by red, pink or white.

Index

Aconite, Winter ... 43
Alison, Small ... 59
Amaranth, Indehiscent ... 79
Amphibious Bistort ... 18
Anemone, Wood ... 45
Archangel, Yellow ... 52
Arrowhead ... 11
Avens, Hybrid ... 46
    Water ... 46
    Wood ... 46
Balsam, Small ... 44
Bedstraw, Fen ... 26
    Marsh ... 26
    Wall ... 94
Betony ... 53
Bindweed, Black ... 72
    Field ... 72
Bird's-foot-trefoil, Greater ... 16
Bitter-cress, Hairy ... 67
    Large ... 21
    Wavy ... 21
Bittersweet ... 29
Bluebell, Common ... 42
    Hybrid ... 42
    Spanish ... 42
Bogbean ... 12
Brooklime ... 27
Brookweed ... 25
Bugle ... 52
Bugloss, Field ... 80
Bur-marigold, Nodding ... 33
    Trifid ... 33
Butterbur ... 32
Buttercup, Celery-leaved ... 16
Byrony, Black ... 50
    White ... 72
Campion, Red ... 49
    White ... 74
Catchfly, Berry ... 49
Catchfly, Night-flowering ... 74
Cat's-ear, Smooth ... 89
Celandine, Greater ... 43
    Lesser ... 43
Chamomile, Corn ... 91
Charlock ... 69
Chickweed, Common ... 73
    Water ... 24
Cinquefoil, Marsh ... 22
Cleavers, Common ... 79
Clover, Strawberry ... 17
Cornflower ... 86
Cornsalad, Common ... 92
Corydalis, Climbing ... 47
    Yellow ... 93
Crane's-bill, Cut-leaved ... 74
Creeping-jenny ... 51
Cress, Hoary ... 65
    Thale ... 67
Cuckooflower ... 22
Cudweed, Broad-leaved ... 89
    Red-tipped ... 55
Dead-nettle, Cut-leaved ... 84
    Henbit ... 84
    Red ... 84
Dock, Clustered ... 23
    Golden ... 23
    Water ... 23
    Wood ... 50
Dog's Mercury ... 41
Dog-violet, Common ... 47
    Early ... 47
Enchanter's-nightshade ... 44
Fat-hen ... 76
Fiddleneck, Common ... 80
Figwort, Water ... 29
    Yellow ... 52
Fleabane, Canadian ... 88
    Guernsey ... 88
Flixweed ... 70
Forget-me-not, Field ... 80
    Tufted ... 27
    Water ... 27
    Wood ... 48
Fumitory, Common ... 62
    Fine-leaved ... 60
Goosefoot, Fig-leaved ... 78
    Maple-leaved ... 78
    Red ... 78
Grass of Parnassus ... 11
Ground-elder ... 92
Groundsel ... 87
Gypsywort ... 30
Helleborine, Broad-leaved ... 40
    Marsh ... 8
Hemlock ... 35
Hemp-agrimony ... 34
Hemp-nettle, Bifid ... 85
    Common ... 85
Herb-robert ... 48
Himalayan Balsam ... 13
Hogweed, Giant ... 36
Iris, Yellow ... 10
Knotgrass, Common ... 71
Lettuce, Wall ... 53
Lily-of-the-valley ... 50
Loosestrife, Purple ... 18
    Yellow ... 16
Lords-and-ladies ... 41
Lousewort, Marsh ... 30
Madder, Field ... 79

Index

Marsh-marigold	14
Mayweed, Scented	91
Scentless	91
Meadow-rue, Common	17
Meadowsweet	17
Melilot, Ribbed	63
White	63
Mercury, Annual	76
Milk-parsley	36
Mint, Corn	83
Water	30
Moschatel	44
Mouse-ear, Sticky	73
Mustard, White	69
Nettle, Small	76
Nightshade,	81
Green	81
Small	81
Orache, Common	77
Grass-leaved	77
Spear-leaved	77
Orchid, Common Spotted	40
Early Marsh	13
Early Purple	40
Heath Spotted	12
Marsh Fragrant	12
Military	38
Pale Marsh	9
Southern Marsh	13
Pansy, Field	63
Parsley, Fool's	92
Pearlwort, Knotted	25
Pellitory-of-the-wall	94
Penny-cress, Field	65
Pennywort, Marsh	29
Pepperwort, Field	65
Perciaria, Pale	71
Phacelia, Tansy-leaved	86
Pimpernel, Bog	26
Scarlet	64
Yellow	51
Pineappleweed	89
Poppy, Common	61
Long-headed	62
Prickly	61
Rough	61
Yellow-juiced	62
Primrose	51
Radish, Wild	68
Ragged-robin	19
Ragwort, Marsh	33
Rape, Oil-seed	69
Redshank	71
Rocket, Hairy	70
Tall	70
Rupturewort, Smooth	54
St. John's-wort, Square-stalked	22
Sandwort, Three-nerved	49
Sanicle	53
Scabious, Devil's-bit	31
Shepherd's-purse	66
Skullcap, Common	31
Soldier, Gallant	87
Shaggy	87
Sow-thistle, Perennial	90
Prickly	90
Smooth	90
Spearwort, Greater	14
Lesser	14
Speedwell, Blue Water	28
Breckland	56
Common Field	82
Fingered	58
Green Field	82
Grey Field	82
Ivy-leaved	83
Marsh	28
Pink Water	28
Spring	57
Wall	83
Wood	48
Spurge, Petty	64
Sun	64
Spurrey, Corn	73
Star-of-Bethlehem, Yellow	39
Stitchwort, Bog	24
Marsh	24
Strawberry, Wild	45
Swine-cress, Common	66
Lesser	66
Thistle, Creeping	88
Marsh	32
Meadow	32
Toadflax, Ivy-leaved	93
Small	86
Twayblade, Common	41
Valerian, Common	34
Marsh	34
Red	94
Venus's-looking-glass	85
Wallflower	93
Wall-rocket, Annual	68
Perennial	68
Water-cress, Common	21
Fool's	35
Water-crowfoot, Common	15
River	15
Thread-leaved	15
Water-dropwort, Fine-leaved	37
River	37